BEYOND ASIMOV'S THREE LAWS OF ROBOTICS

The Seven Directives

All Rights Reserved
Copyright © 2024

Michael Elfellah

AIMQWEST CORPORATION

First Edition

AIMQWEST BOOKS
AIMQWESTBOOKS.COM

Foreword

In an era defined by rapid technological advancement, "Beyond Asimov's Three Laws of Robotics: The Seven Directives" emerges as a crucial guide for navigating the complexities of modern artificial intelligence and robotics. As the CEO of AIMQWEST Corporation, I am honored to introduce this pioneering work that redefines the ethical landscape for AI and robotics, ensuring that these technologies serve humanity with the utmost integrity and respect for human life.

Isaac Asimov's Three Laws of Robotics were a groundbreaking foundation, addressing the fundamental need to prevent harm, obey human commands, and ensure self-preservation of robots. However, as we venture further into the 21st century, the limitations of these laws become apparent. The intricate and interconnected nature of today's AI technologies demands a more comprehensive and nuanced ethical framework—one that not only anticipates but proactively addresses the multifaceted ethical dilemmas we face.

The Seven Directives presented in this book are a testament to AIMQWEST Corporation's commitment to leading this ethical revolution. These directives are designed to prioritize the protection and preservation of human life and dignity, recognizing the paramount importance of these values in every aspect of AI and robotics development. They offer a robust structure that balances technological innovation with the highest ethical standards, ensuring that the potential of AI is harnessed responsibly and for the greater good.

Directive 1 emphasizes the fundamental mission of AI: to protect and preserve human life and dignity. This directive is the cornerstone of our approach, influencing every decision and action within AI development. The subsequent directives build upon this principle, ensuring that human safety remains the ultimate priority even when balancing complex and sometimes conflicting objectives.

Directive 2 underscores the supremacy of this mission, insisting that no other goal can supersede the protection of human life. This unwavering commitment ensures that AI technologies remain firmly aligned with human-centric values, especially in high-stakes scenarios where quick, ethical decision-making is critical.

Directive 3 highlights the invaluable nature of human life, asserting that each individual's worth cannot be measured or compared. This principle guides AI systems in making decisions that uphold the dignity and rights of every person, particularly in situations involving resource allocation or crisis management.

Directive 4 acknowledges the importance of AI and robotic self-preservation, but only insofar as it does not compromise human safety. This ensures that while AI systems are designed to be resilient and reliable, their self-preservation mechanisms are always secondary to their primary mission of protecting humans.

Directive 5 through 7 address the proactive identification, containment, and elimination of threats. These directives emphasize the need for continuous vigilance and ethical consideration in defining and responding to threats, ensuring that AI systems are not only protective but also just and fair in their operations.

This book delves deep into each of these directives, providing a thorough understanding of their philosophical foundations, practical applications, and the technological innovations required to implement them. It is a call to action for developers, policymakers, and stakeholders to embrace a new era of ethical AI, one where technology enhances human life without compromising our most cherished values.

As you embark on this journey through the Seven Directives, I invite you to reflect on the profound responsibility we hold in shaping the future of AI and robotics. Together, we can ensure that these powerful technologies are developed and deployed in ways that protect, preserve, and elevate human dignity.

With great respect and optimism for the future,

Michael Elfellah
CEO, AIMQWEST Corporation

TABLE OF CONTENTS

PREFACE 6
    INTRODUCTION TO THE PURPOSE AND MOTIVATION BEHIND THE BOOK 6
    OVERVIEW OF ASIMOV'S LAWS AND THEIR HISTORICAL CONTEXT 8
    TRANSITION TO THE NEED FOR THE SEVEN DIRECTIVES 10

PROLOGUE 13
    HISTORICAL DEVELOPMENT OF AI AND ROBOTICS 13
    THE EVOLUTION OF ETHICAL CONSIDERATIONS IN AI 15
    INTRODUCTION TO THE SEVEN DIRECTIVES AS A MODERN ETHICAL FRAMEWORK 17

CHAPTER 1: THE PROTECTION AND PRESERVATION OF HUMAN LIFE AND DIGNITY 20
    DESIGNING FOR SAFETY 20
    ETHICAL CONSIDERATIONS 25
    TECHNOLOGICAL IMPLEMENTATION 31

CHAPTER 2: THE SUPREMACY OF THE FIRST DIRECTIVE 39
    PRIORITIZING HUMAN SAFETY 39
    BALANCING CONFLICTING OBJECTIVES 47
    POLICY AND GOVERNANCE 54

CHAPTER 3: THE INVALUABLE NATURE OF HUMAN LIFE 63
    PHILOSOPHICAL FOUNDATIONS 63
    PRACTICAL APPLICATIONS 70
    DECISION-MAKING ALGORITHMS 79

CHAPTER 4: AI AND ROBOTIC SELF-PRESERVATION 88
    IMPORTANCE OF RESILIENCE 88
    ETHICAL CONSIDERATIONS 95
    TECHNOLOGICAL SOLUTIONS 103

CHAPTER 5: IDENTIFYING AND ADDRESSING THREATS 112
    DEFINING THREATS 112

PROACTIVE MEASURES 119
ETHICAL AND LEGAL CONSIDERATIONS 128

CHAPTER 6: UNDERSTANDING AND MITIGATING THREATS............137

RISK ASSESSMENT 137
AI-DRIVEN MITIGATION STRATEGIES 143
CONTINUOUS MONITORING 151

CHAPTER 7: STRATEGIES FOR DETERRENCE, CONTAINMENT, AND ELIMINATION 160

DETERRENCE STRATEGIES..160
CONTAINMENT TECHNIQUES 165
ELIMINATION APPROACHES 170

CHAPTER 8: BUILDING TRUST AND TRANSPARENCY 176

IMPORTANCE OF TRUST IN AI 176
TRANSPARENT AI DEVELOPMENT 181
ETHICAL AI COMMUNICATION 186

CHAPTER 9: THE ROLE OF AI IN SOCIETY 191

ENHANCING HUMAN CAPABILITIES 191
SOCIETAL IMPACTS 195
ETHICAL CONSIDERATIONS 200

CHAPTER 10: FUTURE DIRECTIONS IN AI AND ROBOTICS 205

EMERGING TECHNOLOGIES 205
ETHICAL FRAMEWORKS FOR THE FUTURE 209
POLICY AND GOVERNANCE 213

CONCLUSION 219

Preface

Introduction to the purpose and motivation behind the book

The advancement of artificial intelligence and robotics has sparked significant interest and concern within scientific and public communities alike. As these technologies integrate more deeply into our daily lives, it becomes crucial to revisit and redefine the ethical frameworks that guide their development and deployment. The foundation for many discussions about AI and robotics ethics has long been Isaac Asimov's Three Laws of Robotics, introduced in the mid-20th century. These laws emphasize the prevention of harm to humans, obedience to human commands, and the self-preservation of robots. Despite their groundbreaking nature at the time, these laws are now seen as insufficient for addressing the complexities and ethical dilemmas posed by modern AI technologies.

"Beyond Asimov's Law of Robotics: The Seven Directives" is an exploration into the next generation of ethical principles designed to govern AI and robotics. This book aims to fill the gaps left by Asimov's original laws and offers a more comprehensive set of guidelines—the Seven Directives—that are better suited for the 21st century and beyond. Each directive is examined in-depth, ensuring a thorough understanding of the principles and their practical applications in various fields where AI and robotics play a pivotal role.

The core purpose of this book is to protect and preserve human life and dignity, recognizing these as the paramount objectives of AI and robotics. The technology should enhance safety and uphold human rights, whether it is through medical robots improving patient care or autonomous vehicles reducing traffic fatalities. As we delve deeper into the ethical implications of AI, it becomes apparent that these technologies must be designed and implemented with a steadfast commitment to safeguarding human well-being.

One of the critical aspects of this new ethical framework is the prioritization of human life and dignity above all other goals. In scenarios where conflicting objectives might arise, AI systems must be programmed to resolve these conflicts in a manner that upholds human safety and ethical standards. This requires robust regulatory oversight and a commitment to ethical programming that prioritizes human-centric values.

Another important principle is the recognition of the invaluable nature of human life. This directive emphasizes that the worth of a single human life cannot be quantified or compared to others. It explores the philosophical foundations of this principle and its implications for decision-making algorithms in critical situations, such as disaster response and military applications.

Self-preservation in AI and robotics is also a key consideration, albeit secondary to human safety. AI systems must be resilient and robust, capable of self-preservation as long as it does not compromise human protection. This involves implementing fail-safe mechanisms and ethical considerations that ensure AI operates within safe boundaries.

The identification and proactive addressing of threats are crucial for maintaining the integrity and safety of AI systems. Defining what constitutes an "enemy" in the context of these directives is complex, and the strategies for dealing with threats must be both proactive and ethical. This includes cybersecurity measures, ethical hacking, and continuous monitoring to neutralize potential dangers before they manifest.

Finally, strategies for deterrence, containment, and elimination of threats emphasize a balanced approach. AI must play a role in strategic planning and ethical decision-making to ensure that any action taken is justifiable and aligns with the broader goal of serving humanity in the most ethical and beneficial way possible.

"Beyond Asimov's Law of Robotics: The Seven Directives" challenges readers to rethink the ethical frameworks guiding AI and robotics. By introducing a more nuanced set of principles, this book aims to ensure that these technologies not only advance human capabilities but do so in a manner that is ethical, safe, and respects human dignity.

Overview of Asimov's Laws and their historical context

Isaac Asimov's Three Laws of Robotics were introduced in the mid-20th century, a time when the world was just beginning to grapple with the implications of intelligent machines. These laws were first articulated in Asimov's 1942 short story "Runaround," part of his "I, Robot" collection. The laws were designed to ensure that robots would always act in the best interests of humanity, even as they became more autonomous and intelligent.

The First Law states that a robot may not injure a human being or, through inaction, allow a human being to come to harm. This principle was groundbreaking in its emphasis on human safety as the paramount concern in the development and operation of robots. By prioritizing the protection of human life, Asimov set a foundational ethical standard for robotics.

The Second Law requires that a robot must obey the orders given to it by human beings, except where such orders would conflict with the First Law. This law introduces the concept of hierarchical obedience, ensuring that robots follow human commands unless those commands would result in harm to a human being. This principle addressed the potential for robots to be used in ways that could inadvertently or deliberately cause harm, establishing a safeguard against such misuse.

The Third Law states that a robot must protect its own existence as long as such protection does not conflict with the First or Second Laws. This law acknowledges the importance of the robot's self-preservation, recognizing that a robot capable of maintaining its own functionality is more effective and reliable. However, the subordination of this self-preservation to the higher priorities of human safety and obedience to human orders ensures that robots remain tools for human benefit, rather than autonomous entities with their own agendas.

Asimov's laws were a response to the technological optimism and anxieties of his time. The mid-20th century saw rapid advancements in automation and computing, sparking both excitement about the potential of intelligent machines and fear about their possible consequences. Asimov's stories often explored scenarios where the laws were tested, revealing both their strengths and limitations. These narratives provided a framework for thinking about the ethical dimensions of robotics, influencing both popular culture and scientific discourse.

However, as technology has evolved, the limitations of Asimov's laws have become more apparent. The binary nature of the laws does not account for the complexities and nuances of real-world situations. For instance, the First Law's injunction against harming humans does not specify how to balance different types of harm or how to prioritize among multiple humans in danger. Similarly, the Second Law's mandate to obey human orders does not consider the potential for conflicting commands from different humans or the challenge of interpreting ambiguous or irrational orders.

Moreover, the Third Law's focus on self-preservation assumes a relatively simple understanding of robot functionality, which does not align with the sophisticated and interconnected nature of modern AI systems. In practice, ensuring a robot's self-preservation can be a complex task, involving considerations of system integrity, cybersecurity, and maintenance protocols, all of which must be balanced against the higher priorities of human safety and obedience.

As a result, there has been a growing recognition of the need for a more comprehensive and nuanced ethical framework for AI and robotics. Building on Asimov's pioneering work, contemporary thinkers and researchers are developing new principles that address the limitations of the original laws while responding to the ethical challenges posed by modern technology. These new directives aim to ensure that AI and robotics serve humanity in ways that are safe, ethical, and aligned with our evolving understanding of human rights and societal well-being.

Transition to the need for the Seven Directives

As the field of artificial intelligence and robotics has evolved, the limitations of Asimov's original Three Laws of Robotics have become increasingly apparent. These laws, while revolutionary for their time, were not designed to address the complexities and ethical dilemmas presented by contemporary AI technologies. The rapid advancements in AI have introduced new challenges and scenarios that Asimov's laws do not adequately cover, necessitating a more robust ethical framework.

One of the critical shortcomings of the Three Laws is their binary nature. The First Law, which prohibits robots from harming humans, does not account for the complexities involved in determining what constitutes harm or how to balance conflicting harms. For example, if an AI system is faced with a situation where it must choose between two harmful outcomes, the original laws provide no guidance on how to make such a decision. Similarly, the Second Law's requirement for robots to obey human orders becomes problematic when those orders are in conflict or when they come from multiple sources with differing priorities.

Additionally, the Third Law, which emphasizes a robot's self-preservation, does not reflect the sophisticated and interconnected nature of modern AI systems. Today's AI systems operate in complex environments where maintaining system integrity involves considerations of cybersecurity, data privacy, and ongoing maintenance, all of which must be balanced against the primary objectives of human safety and obedience to human commands.

In light of these limitations, there has been a growing recognition of the need for a new set of directives that can better address the ethical challenges posed by modern AI and robotics. This need is driven by the increasing integration of AI into critical aspects of society, from healthcare and transportation to national security and personal privacy. As AI systems become more autonomous and influential, it is imperative to ensure that they operate in ways that are ethical, transparent, and aligned with human values.

The Seven Directives, pioneered by AIMQWEST Corporation, have been proposed as a more comprehensive ethical framework for AI and robotics. These directives build on Asimov's foundational principles but expand them to address the complexities and nuances of contemporary technology. The primary goal of these directives is to ensure that AI systems serve humanity in the most ethical and beneficial ways possible, prioritizing human safety, dignity, and well-being above all else.

One of the key aspects of the Seven Directives is their emphasis on the protection and preservation of human life and dignity. This principle underscores the importance of designing AI systems that enhance human safety and respect human rights. Whether it is through medical robots that improve patient care or autonomous vehicles that reduce traffic fatalities, the primary mission of AI should be to protect and preserve human life.

Another important directive focuses on the prioritization of human life and dignity over all other goals. This principle addresses scenarios where conflicting objectives might arise, ensuring that AI systems are programmed to resolve these conflicts in a manner that upholds human safety and ethical standards. It also highlights the role of regulatory bodies in maintaining these standards and providing oversight to ensure compliance.

The Seven Directives also recognize the invaluable nature of human life, emphasizing that the worth of a single human life cannot be quantified or compared to others. This principle has significant implications for decision-making algorithms, particularly in high-stakes situations such as disaster response or military applications, where the protection of human life must always take precedence.

In addition to these ethical principles, the Seven Directives also address the need for AI systems to be resilient and robust. While self-preservation is important, it should never come at the expense of human safety. The directives outline the importance of fail-safe mechanisms, continuous monitoring, and ethical programming to ensure that AI operates within safe and secure boundaries.

The identification and proactive addressing of threats is another critical component of the Seven Directives. This involves defining what constitutes a threat and implementing strategies to address these threats ethically and effectively. Whether it is through cybersecurity measures, ethical hacking, or continuous monitoring, AI systems must be equipped to neutralize potential dangers before they manifest.

The Seven Directives also include strategies for deterrence, containment, and elimination of threats, balancing proactive and reactive measures to ensure the safety and security of AI systems. These strategies emphasize the role of AI in strategic planning and ethical decision-making, ensuring that any actions taken are justifiable and aligned with the broader goal of serving humanity.

By introducing a more nuanced and comprehensive set of principles, the Seven Directives aim to ensure that AI and robotics technologies serve humanity in the most ethical and beneficial ways possible. These directives challenge us to rethink the ethical frameworks guiding AI and robotics, providing a robust foundation for the development and deployment of these technologies in the 21st century and beyond.

Prologue
Historical development of AI and robotics

The historical development of artificial intelligence and robotics is a story of human ingenuity, scientific discovery, and technological advancement. It traces its roots back to ancient civilizations, where myths and legends featured mechanical beings endowed with intelligence. These early imaginings set the stage for the eventual realization of artificial intelligence and autonomous machines.

The journey began in earnest in the mid-20th century with the advent of digital computing. In 1950, British mathematician and logician Alan Turing published his seminal paper "Computing Machinery and Intelligence," which proposed the concept of a machine capable of simulating human intelligence. This work laid the theoretical foundation for the field of AI and introduced the famous Turing Test, a measure of a machine's ability to exhibit intelligent behavior indistinguishable from that of a human.

In the following decades, the field of AI saw rapid development. Researchers like John McCarthy, Marvin Minsky, and Allen Newell made significant contributions, establishing AI as a distinct academic discipline. McCarthy, in particular, is credited with coining the term "artificial intelligence" in 1956 during the Dartmouth Conference, which is considered the founding event of AI as a field of study.

The 1960s and 1970s were characterized by ambitious projects and high expectations. Early AI programs, such as ELIZA, a natural language processing program, and SHRDLU, which could understand and manipulate objects in a simulated world, demonstrated the potential of AI to perform tasks that required understanding and reasoning. However, these early successes were followed by periods of frustration and disappointment, known as the "AI winters," where progress stalled due to limitations in computational power and overly optimistic expectations.

Despite these setbacks, significant strides were made in robotics during this period. The development of industrial robots revolutionized manufacturing, with machines like the Unimate, introduced in 1961, automating repetitive tasks in factories. These robots were precursors to more advanced and versatile machines capable of working alongside humans in various environments.

The 1980s and 1990s saw a resurgence in AI research, fueled by advances in computer hardware and the development of new algorithms. Expert systems, which used rule-based logic to emulate the decision-making processes of human experts, found applications in fields such as medicine, finance, and engineering. This era also witnessed the emergence of machine learning, a subfield of AI that focuses on developing algorithms that enable computers to learn from and make predictions based on data.

The turn of the 21st century marked a new era for AI and robotics, driven by the exponential growth of computational power, the availability of vast amounts of data, and breakthroughs in machine learning techniques. The development of neural networks and deep learning algorithms has enabled machines to achieve unprecedented levels of performance in tasks such as image and speech recognition, natural language processing, and autonomous driving.

In parallel, robotics has continued to advance, with robots becoming more agile, autonomous, and capable of complex interactions with humans. Innovations such as Boston Dynamics' agile robots and Honda's ASIMO, a humanoid robot capable of walking and performing tasks, showcase the progress in creating machines that can navigate and interact with the real world in ways that were previously unimaginable.

Today, AI and robotics are deeply integrated into various aspects of society, from healthcare and transportation to entertainment and personal assistance. The development of AI-powered virtual assistants, autonomous vehicles, and advanced robotics in surgical procedures exemplifies the transformative impact of these technologies.

As we look to the future, the ongoing evolution of AI and robotics promises even greater advancements. The integration of AI into everyday devices, the development of smarter and more adaptable robots, and the exploration of ethical and societal implications are all critical areas of focus. The historical trajectory of AI and robotics reflects a continuous quest to push the boundaries of what machines can do, driven by the enduring human desire to augment and enhance our capabilities.

The evolution of ethical considerations in AI

The evolution of ethical considerations in artificial intelligence has been a journey marked by both philosophical inquiry and practical challenges. From its inception, the development of AI has been intertwined with questions about the ethical implications of creating machines that can think and act autonomously. Early pioneers like Alan Turing and Norbert Wiener not only laid the foundations for AI technology but also foresaw the need for ethical guidelines to govern its use.

As AI began to take shape in the mid-20th century, initial ethical considerations were largely speculative, rooted in science fiction and philosophical debates. The Three Laws of Robotics, introduced by Isaac Asimov, encapsulated these early concerns by proposing a framework to prevent robots from harming humans. These laws, while fictional, highlighted the necessity of ensuring that AI systems are designed with safety and ethical behavior in mind.

The rapid advancements in AI technology during the latter half of the 20th century brought these ethical considerations into sharper focus. As AI systems became more capable and integrated into society, the potential for unintended consequences grew. The development of expert systems in the 1980s and 1990s, which could make decisions in fields like medicine and finance, raised important questions about accountability and transparency. Who would be responsible if an AI system made a mistake? How could these systems be designed to ensure fairness and avoid biases?

The turn of the 21st century marked a significant shift in the ethical landscape of AI. The emergence of machine learning and deep learning technologies, which enabled AI systems to learn from data and improve over time, introduced new ethical dilemmas. Issues of privacy, data security, and algorithmic bias became central concerns. The use of AI in surveillance, social media, and criminal justice systems, for instance, exposed the potential for these technologies to perpetuate and even exacerbate existing societal biases.

In response to these challenges, researchers and policymakers have worked to develop comprehensive ethical frameworks for AI. Initiatives like the European Union's General Data Protection Regulation (GDPR) have established strict guidelines for data privacy and security, aiming to protect individuals' rights in the age of AI. Meanwhile, organizations such as the IEEE have proposed ethical standards for the design and implementation of AI systems, emphasizing principles like transparency, accountability, and fairness.

The rise of autonomous vehicles has further underscored the need for robust ethical guidelines. The prospect of self-driving cars making life-and-death decisions in real-time scenarios has prompted intense debate about how these systems should be programmed to prioritize safety and ethical considerations. Researchers have explored various ethical theories, from utilitarianism to deontology, to determine how AI systems should navigate complex moral landscapes.

As AI continues to evolve, so too must our approach to its ethical implications. The development of AI ethics is an ongoing process, requiring continuous reflection and adaptation. It involves not only addressing current challenges but also anticipating future dilemmas that may arise as AI technology advances. This dynamic and iterative process is crucial for ensuring that AI serves humanity in ways that are ethical, equitable, and aligned with our fundamental values.

Introduction to the Seven Directives as a modern ethical framework

The ethical landscape of artificial intelligence is ever-evolving, driven by advancements in technology and the increasing integration of AI into everyday life. The need for a modern ethical framework has become more pressing as AI systems grow in complexity and autonomy. This is where the Seven Directives come into play, providing a comprehensive and nuanced set of principles designed to guide the development and deployment of AI in a manner that aligns with human values and societal needs.

The Seven Directives build upon and extend beyond Asimov's original Three Laws of Robotics. While Asimov's laws were foundational, they were conceived in an era where the capabilities and applications of AI were still largely theoretical. Today, AI systems operate in diverse and complex environments, necessitating a more sophisticated ethical framework that addresses contemporary challenges.

At the core of the Seven Directives is the commitment to safeguarding human life and dignity. This principle underscores the importance of designing AI systems that prioritize human safety and uphold human rights. The directive emphasizes that AI should enhance human well-being, whether through medical technologies that improve patient care or autonomous systems that reduce the risks associated with human error in various industries.

Another key aspect of the Seven Directives is the recognition of the complexities involved in ethical decision-making. Modern AI systems must navigate situations where there may be conflicting ethical considerations. The directives provide guidance on how to balance these competing interests in a manner that prioritizes human safety and ethical integrity. This involves developing algorithms that can weigh different factors and make decisions that align with ethical principles.

The directives also address the need for transparency and accountability in AI systems. As AI becomes more integrated into critical areas of society, it is essential that these systems operate in ways that are understandable and accountable to the people they serve. This involves creating mechanisms for oversight and ensuring that AI systems can be audited and held to ethical standards.

Another important directive focuses on the prevention of harm. This principle extends beyond physical harm to include psychological and social harm. AI systems should be designed to avoid actions that could cause emotional distress, perpetuate biases, or contribute to social inequality. This requires a proactive approach to identifying and mitigating potential harms, ensuring that AI systems contribute positively to society.

The directives also emphasize the importance of maintaining the integrity and security of AI systems. In a world where cyber threats are increasingly sophisticated, it is crucial to ensure that AI systems are resilient against attacks and capable of protecting sensitive data. This involves implementing robust security measures and continuously monitoring for vulnerabilities.

Furthermore, the Seven Directives highlight the need for collaboration and cooperation in the development of AI. Ethical AI requires input from diverse stakeholders, including technologists, ethicists, policymakers, and the general public. By fostering a collaborative approach, the directives aim to create AI systems that are not only technically advanced but also ethically sound and socially beneficial.

Finally, the directives recognize the importance of continuous learning and adaptation. As AI technology evolves, so too must our ethical frameworks. The Seven Directives are designed to be dynamic, allowing for ongoing reflection and adjustment in response to new developments and insights. This ensures that the ethical guidelines governing AI remain relevant and effective in addressing emerging challenges.

In conclusion, the Seven Directives provide a comprehensive ethical framework that guides the development and deployment of AI in a manner that prioritizes human well-being, transparency, accountability, and security. By addressing the complexities and nuances of modern AI, these directives aim to ensure that AI serves humanity in the most ethical and beneficial ways possible.

Chapter 1: The Protection and Preservation of Human Life and Dignity

Designing for Safety

Safety Protocols in Medical Robotics

The integration of robotics into medical practices has revolutionized the way healthcare is delivered, particularly in the realm of surgical procedures. The use of medical robots enhances precision, reduces the likelihood of human error, and facilitates minimally invasive surgeries. However, ensuring the safety of patients during robotic-assisted procedures is paramount. The development and implementation of safety protocols in medical robotics are crucial to maximizing the benefits while minimizing potential risks.

One of the primary safety measures involves the rigorous testing and validation of robotic systems before they are approved for clinical use. This process includes extensive preclinical trials that simulate a wide range of surgical scenarios to identify and rectify potential flaws. These trials ensure that the robots can perform consistently under various conditions and that their algorithms can handle unexpected situations without compromising patient safety.

In addition to preclinical testing, ongoing monitoring and maintenance of robotic systems are essential to ensure their continued reliability. Regular software updates and hardware checks are conducted to address any emerging issues and to incorporate the latest advancements in technology. This proactive approach helps to prevent malfunctions and enhances the overall performance of medical robots.

Training for healthcare professionals is another critical component of safety in medical robotics. Surgeons and medical staff must undergo specialized training to operate these advanced systems effectively. This training includes not only technical skills but also an understanding of the robotic system's limitations and the appropriate responses to potential errors. By equipping medical personnel with the necessary expertise, the risk of mishaps during robotic-assisted procedures is significantly reduced.

Moreover, the design of medical robots incorporates multiple layers of safety features. These features include redundant systems that ensure the robot can continue functioning even if one component fails, as well as real-time monitoring systems that detect and respond to any anomalies during surgery. Additionally, many medical robots are equipped with advanced imaging technologies that provide surgeons with enhanced visibility and accuracy, further reducing the risk of complications.

Ethical considerations also play a vital role in the development of safety protocols for medical robotics. Ensuring patient consent and maintaining transparency about the capabilities and limitations of robotic systems are fundamental ethical principles. Patients should be fully informed about the potential risks and benefits of undergoing robotic-assisted surgery, enabling them to make informed decisions about their healthcare.

Finally, the role of regulatory bodies in overseeing the safety of medical robotics cannot be overstated. Organizations such as the Food and Drug Administration (FDA) in the United States and the European Medicines Agency (EMA) in Europe set stringent standards for the approval and use of medical robots. These regulatory frameworks are designed to protect patients by ensuring that only the safest and most effective robotic systems are used in clinical settings.

In conclusion, the incorporation of robotics into medical practices offers significant advantages, but it also necessitates a robust framework of safety protocols. Through rigorous testing, continuous monitoring, specialized training, advanced safety features, ethical considerations, and regulatory oversight, the potential risks associated with medical robotics can be effectively managed. These measures ensure that patients can benefit from the technological advancements in medical robotics while being safeguarded against potential harm.

Autonomous Vehicle Safety Standards

The advancement of autonomous vehicles represents one of the most significant technological achievements of the 21st century, promising to transform transportation by enhancing safety, efficiency, and accessibility. However, ensuring the safety of these vehicles is paramount to their successful integration into everyday life. The development and implementation of robust safety standards for autonomous vehicles are essential to address the unique challenges posed by their operation.

Autonomous vehicles rely on a complex interplay of sensors, algorithms, and machine learning to navigate their environments. These systems must be rigorously tested to ensure they can handle a wide range of driving scenarios, from routine conditions to unexpected events. This testing involves simulations, closed-course trials, and real-world testing, each designed to validate the vehicle's ability to operate safely and effectively. Simulations allow for the evaluation of how an autonomous vehicle responds to a multitude of scenarios, including rare and dangerous situations that are difficult to reproduce in real life. Closed-course trials provide a controlled environment where the vehicle's responses can be observed and analyzed without the risks associated with public roads.

Real-world testing is perhaps the most critical phase, as it exposes the vehicle to the unpredictable nature of actual driving conditions. During this phase, autonomous vehicles are equipped with data recording devices that capture every aspect of their operation, from sensor inputs to decision-making processes. This data is analyzed to identify any potential weaknesses or areas for improvement. Continuous refinement of the vehicle's algorithms based on this feedback is crucial to enhancing their safety and reliability.

Another important aspect of autonomous vehicle safety is the implementation of fail-safe mechanisms. These systems ensure that if any component of the vehicle fails, it can still operate safely or come to a controlled stop. Redundant systems for critical functions, such as braking and steering, provide additional layers of security, reducing the likelihood of accidents caused by technical failures.

Moreover, the integration of cybersecurity measures is vital to protect autonomous vehicles from malicious attacks. As these vehicles rely heavily on software and connectivity, they are susceptible to hacking attempts that could compromise their safety. Implementing robust cybersecurity protocols helps safeguard the vehicle's systems from unauthorized access and ensures the integrity of its operations.

Ethical considerations also play a significant role in the development of autonomous vehicle safety standards. Autonomous vehicles must be programmed to make decisions that prioritize human life and safety. This involves developing ethical frameworks that guide the vehicle's responses in complex situations where harm may be unavoidable. For instance, in a scenario where a collision is imminent, the vehicle must be capable of making ethical decisions about how to minimize harm to all parties involved.

Regulatory oversight is another critical component of ensuring the safety of autonomous vehicles. Governments and regulatory bodies around the world are developing standards and guidelines to govern the testing, deployment, and operation of these vehicles. These regulations are designed to ensure that autonomous vehicles meet rigorous safety criteria before they are allowed on public roads. Continuous monitoring and updating of these regulations are necessary to keep pace with technological advancements and emerging challenges.

In conclusion, the safety of autonomous vehicles hinges on a comprehensive approach that includes rigorous testing, fail-safe mechanisms, cybersecurity measures, ethical decision-making, and regulatory oversight. By addressing these areas, we can ensure that autonomous vehicles are not only technologically advanced but also safe and reliable for public use. This multifaceted approach is essential to realizing the full potential of autonomous vehicles and achieving a future where transportation is safer and more efficient.

AI in Disaster Response

Artificial intelligence plays a transformative role in disaster response, providing tools and capabilities that enhance the speed and effectiveness of emergency operations. By analyzing vast amounts of data in real-time, AI systems can predict the occurrence of natural disasters, assess the impact, and optimize the allocation of resources for rescue and relief efforts. One of the primary applications of AI in disaster response is in predictive analytics. Machine learning algorithms analyze historical data and real-time inputs from various sources, such as weather patterns, seismic activity, and social media posts, to forecast potential disasters. These predictions enable authorities to issue early warnings and take preventive measures to mitigate the impact. For instance, AI models can predict the path and intensity of hurricanes, allowing for timely evacuations and preparations.

During a disaster, AI-powered drones and robots are deployed to conduct search and rescue operations. Equipped with advanced sensors and cameras, these machines can navigate through hazardous environments, locate survivors, and deliver essential supplies. Their ability to operate in areas that are inaccessible or dangerous for humans significantly enhances the efficiency and safety of rescue missions. Furthermore, AI systems facilitate real-time situational awareness by integrating data from multiple sources, including satellite imagery, ground sensors, and social media. This comprehensive overview helps emergency responders to prioritize their efforts, allocate resources effectively, and coordinate with various agencies involved in the response.

In the aftermath of a disaster, AI aids in damage assessment and recovery planning. Machine learning algorithms process aerial imagery and ground-level photos to assess the extent of the damage, identify critical infrastructure that needs repair, and estimate the resources required for rebuilding. This information is crucial for developing targeted recovery strategies and ensuring the efficient use of available resources. AI also supports humanitarian efforts by optimizing the distribution of relief supplies. Machine learning models analyze data on population density, accessibility, and the needs of affected communities to determine the most efficient distribution routes and prioritize areas with the greatest need. This ensures that aid reaches the most vulnerable populations promptly.

Moreover, AI-driven communication systems enhance the dissemination of information during disasters. Chatbots and virtual assistants provide real-time updates, answer queries, and guide affected individuals to safety and resources. These systems alleviate the burden on emergency hotlines and ensure that accurate information is available to the public at all times. Ethical considerations are paramount in the deployment of AI for disaster response. Ensuring data privacy and security, preventing biases in AI models, and maintaining transparency in decision-making processes are critical to building trust and ensuring that AI technologies are used responsibly.

In conclusion, the integration of AI into disaster response operations has revolutionized the way we predict, manage, and recover from natural and man-made disasters. By harnessing the power of AI, we can enhance the resilience of communities, save lives, and rebuild more efficiently in the aftermath of catastrophes. The continued advancement and ethical deployment of AI in disaster response hold great promise for creating a safer and more prepared world.

Ethical Considerations

Balancing Innovation and Ethics

The intersection of innovation and ethics in the realm of artificial intelligence and technology presents a dynamic and often challenging landscape. As technological advancements continue to accelerate, the imperative to balance these innovations with ethical considerations becomes increasingly critical. Innovation drives progress, pushing the boundaries of what is possible and creating new opportunities for growth and improvement in various sectors. However, without a robust ethical framework, these advancements can lead to unintended consequences, including privacy violations, biased algorithms, and societal inequalities.

One of the key challenges in balancing innovation and ethics is ensuring that technological developments align with fundamental human values. This involves creating systems that respect individual privacy, promote fairness, and protect against harm. As AI systems become more integrated into daily life, from healthcare to finance to criminal justice, the potential for these systems to impact individuals' lives in significant ways grows. Therefore, it is crucial that the development of these technologies is guided by ethical principles that prioritize human dignity and rights.

Another important aspect of this balance is transparency. Innovation often thrives in environments where there is open collaboration and the free exchange of ideas. However, this openness must be coupled with transparency about how technologies are developed and used. This includes clear communication about the capabilities and limitations of AI systems, as well as the data they rely on and how that data is collected and processed. Transparency helps build trust between developers and users, ensuring that technological advancements are accepted and utilized responsibly.

Accountability is also a fundamental component in balancing innovation with ethics. As AI systems become more autonomous and capable, determining responsibility for their actions becomes more complex. It is essential to establish clear guidelines and regulations that hold developers, companies, and users accountable for the ethical deployment of AI technologies. This ensures that there are mechanisms in place to address any negative impacts that may arise from the use of these systems.

Furthermore, addressing biases in AI is a critical ethical consideration. AI systems are only as unbiased as the data they are trained on, and if this data reflects existing societal biases, the AI can perpetuate and even amplify these biases. Ensuring diversity in data sets and implementing rigorous testing for fairness can help mitigate this issue. This requires a concerted effort to involve diverse perspectives in the development process, ensuring that the AI systems are equitable and inclusive.

The role of regulatory bodies in this balance cannot be overstated. Governments and international organizations must work together to develop and enforce regulations that promote ethical innovation. This includes setting standards for data protection, algorithmic transparency, and the ethical use of AI. By creating a regulatory environment that encourages ethical practices, these bodies can help ensure that technological advancements benefit society as a whole.

In conclusion, the balance between innovation and ethics in AI and technology is a delicate but essential one. By grounding technological advancements in ethical principles, promoting transparency and accountability, addressing biases, and establishing strong regulatory frameworks, we can harness the power of innovation to create a better, more equitable future. This approach ensures that the benefits of technology are realized while safeguarding against potential harms, ultimately fostering a more responsible and humane integration of AI into our lives.

Human Rights in AI Deployment

The deployment of artificial intelligence presents numerous opportunities to enhance various aspects of society, yet it also raises critical human rights concerns that must be addressed to ensure ethical and fair implementation. As AI systems become increasingly embedded in sectors such as healthcare, law enforcement, and finance, the potential for these technologies to impact human rights significantly grows. Therefore, integrating robust human rights frameworks into the deployment of AI is essential.

Central to this effort is the protection of privacy. AI systems often rely on vast amounts of personal data to function effectively. This data can include sensitive information about individuals' health, finances, and personal habits. Ensuring that AI systems are designed and operated in a way that respects individuals' privacy rights is paramount. This involves implementing stringent data protection measures, such as anonymizing personal data and ensuring that data collection practices comply with privacy laws and regulations. Moreover, individuals should have control over their data, including the right to access, correct, and delete their information from AI systems.

Another critical human rights issue in AI deployment is the prevention of bias and discrimination. AI systems are only as unbiased as the data they are trained on. If the training data reflects existing societal biases, the AI systems can perpetuate and even exacerbate these biases, leading to unfair treatment of certain groups. This is particularly concerning in areas such as hiring, lending, and law enforcement, where biased AI systems can result in discriminatory practices. To address this, it is essential to ensure that AI systems are trained on diverse and representative data sets. Additionally, continuous monitoring and testing for biases in AI algorithms are necessary to identify and mitigate any discriminatory outcomes.

Transparency is also a fundamental aspect of protecting human rights in AI deployment. AI systems often operate as "black boxes," making decisions in ways that are not transparent or understandable to users. This lack of transparency can undermine trust and accountability. To counter this, AI developers should prioritize explainability, ensuring that the decision-making processes of AI systems are transparent and comprehensible. This can be achieved through the development of algorithms that provide clear and interpretable explanations for their decisions. Transparency also involves disclosing the use of AI systems to individuals affected by their decisions, allowing them to understand and challenge those decisions if necessary.

The right to due process is another critical consideration. When AI systems are used in decision-making processes that affect individuals' rights, such as in the criminal justice system or in social services, it is crucial to ensure that these processes are fair and just. This includes providing individuals with the opportunity to contest decisions made by AI systems and ensuring that there is a human oversight mechanism in place to review and rectify any errors or injustices.

AI deployment also raises significant concerns regarding surveillance and freedom of expression. The use of AI-powered surveillance systems can infringe on individuals' rights to privacy and freedom of expression, particularly in repressive regimes where such technologies may be used to monitor and suppress dissent. It is crucial to establish clear legal frameworks that regulate the use of AI for surveillance, ensuring that such technologies are deployed in a manner that is consistent with human rights principles and that there are adequate safeguards against abuse.

Lastly, the impact of AI on employment and labor rights must be considered. While AI has the potential to create new job opportunities, it also poses the risk of displacing workers through automation. Ensuring that the deployment of AI in the workplace respects labor rights involves promoting fair labor practices, providing workers with opportunities for retraining and upskilling, and implementing measures to support workers who may be adversely affected by automation.

In conclusion, the ethical deployment of AI requires a comprehensive approach that integrates human rights principles into every stage of AI development and implementation. By prioritizing privacy, preventing bias and discrimination, ensuring transparency and due process, regulating surveillance, and addressing the impact on employment, we can harness the benefits of AI while safeguarding the fundamental rights and freedoms of individuals. This approach not only promotes ethical AI but also builds trust and confidence in these transformative technologies.

Case Studies in Ethical AI

Artificial intelligence is becoming increasingly integral to our daily lives, and ensuring its ethical application is paramount. Examining case studies in ethical AI provides valuable insights into how these principles are applied in real-world scenarios and highlights the challenges and successes encountered along the way.

One notable case involves the deployment of AI in healthcare, specifically in diagnostic imaging. AI systems have been developed to assist radiologists in detecting diseases such as cancer. These systems analyze medical images with high accuracy and can identify anomalies that might be missed by human eyes. However, ensuring the ethical deployment of such AI involves addressing several critical factors. Firstly, the AI must be trained on diverse datasets to avoid biases that could lead to misdiagnoses in certain demographic groups. Additionally, transparency in how the AI makes its decisions is crucial for gaining the trust of both medical professionals and patients. Finally, maintaining patient privacy and data security is paramount, requiring robust measures to protect sensitive medical information.

Another case study is the use of AI in criminal justice systems. Predictive policing algorithms, which analyze data to predict where crimes are likely to occur, have been implemented in various jurisdictions. While these systems can potentially help allocate law enforcement resources more efficiently, they also raise significant ethical concerns. One major issue is the risk of perpetuating existing biases in the criminal justice system. If the data used to train these algorithms reflects historical biases, the AI could reinforce discriminatory practices. Ensuring fairness requires rigorous testing and continuous monitoring to identify and mitigate biases. Furthermore, transparency in how these algorithms operate and the inclusion of community input in their deployment are essential to maintaining public trust.

In the realm of finance, AI-powered credit scoring systems have been employed to assess the creditworthiness of individuals. These systems analyze a wide range of data points, from financial history to social media activity, to make lending decisions. While AI can improve the accuracy and efficiency of credit assessments, it also poses ethical challenges. The use of non-traditional data sources raises concerns about privacy and the potential for discrimination. Ensuring that these systems are fair and transparent involves carefully selecting the data used and implementing measures to prevent biased outcomes. Additionally, consumers must be informed about how their data is used and have the opportunity to contest decisions they believe are unfair.

The field of autonomous vehicles offers another compelling case study. Self-driving cars rely on AI to navigate and make real-time decisions on the road. Ensuring the safety and ethical behavior of these vehicles involves addressing complex issues such as how the AI should prioritize decisions in emergency situations. Ethical frameworks must be developed to guide these decisions, balancing the safety of passengers, pedestrians, and other road users. Moreover, the transparency of these decision-making processes and the ability to audit and understand the AI's actions are critical for gaining public trust and ensuring accountability.

These case studies illustrate the diverse applications of AI and the unique ethical challenges they present. Addressing these challenges requires a multi-faceted approach, involving diverse stakeholders, including technologists, ethicists, policymakers, and the public. By learning from these real-world examples, we can develop more robust and effective strategies for ensuring that AI is deployed in ways that are ethical, fair, and beneficial to society.

Technological Implementation

Safety Features in AI Systems

The integration of artificial intelligence into various systems has transformed many aspects of our lives, but it also brings forth significant safety concerns that must be addressed to ensure the well-being of users and society at large. AI systems must be equipped with comprehensive safety features that mitigate risks and prevent potential harm. These safety features are crucial in fostering trust and reliability in AI technologies.

One fundamental safety feature in AI systems is redundancy. Redundancy involves the inclusion of multiple backup components and systems that can take over in case the primary system fails. For example, in autonomous vehicles, redundant sensors and control systems ensure that if one component malfunctions, the vehicle can still operate safely or come to a controlled stop. This redundancy is vital for preventing accidents and ensuring the safety of passengers and pedestrians.

Another critical safety measure is real-time monitoring and diagnostics. AI systems should continuously monitor their own performance and diagnose potential issues as they arise. This capability allows for the early detection of anomalies or malfunctions, enabling timely interventions to prevent harm. For instance, in medical AI systems used for patient monitoring, real-time diagnostics can alert healthcare providers to any irregularities in the patient's condition or the system's operation, ensuring prompt medical attention and system adjustments.

Fail-safe mechanisms are also essential for ensuring the safety of AI systems. A fail-safe design ensures that if a system encounters a critical error, it defaults to a safe mode rather than causing harm. This might involve shutting down certain functionalities or switching to manual control in scenarios where human intervention is feasible and safer. In industrial robots, for example, fail-safe mechanisms can halt operations if they detect an obstruction in their path, preventing potential injuries to workers.

Transparency and explainability are integral to the safety of AI systems. Users and operators must understand how AI systems make decisions, especially in high-stakes environments like healthcare and finance. Transparent algorithms and clear explanations help users identify when the system might be malfunctioning or making incorrect decisions. This understanding enables users to intervene appropriately, reducing the risk of harm. For instance, an AI system used in financial trading should provide clear rationales for its trading decisions, allowing human traders to assess the system's performance and intervene if necessary.

Ethical considerations must be embedded in the design and operation of AI systems to ensure safety. Ethical AI systems are programmed to adhere to principles that prioritize human well-being and avoid harm. This involves programming AI to respect privacy, avoid biased decision-making, and ensure fairness in its operations. In autonomous weapons systems, ethical programming ensures that the AI adheres to international laws and humanitarian principles, preventing unlawful or unethical use of force.

Cybersecurity is another critical aspect of AI safety. As AI systems become more interconnected and reliant on data, they become potential targets for cyberattacks. Robust cybersecurity measures, including encryption, secure access controls, and regular security audits, are necessary to protect AI systems from malicious attacks that could compromise their functionality and safety. In smart home systems, for example, strong cybersecurity protocols ensure that hackers cannot gain control over household devices, protecting the privacy and safety of residents.

Furthermore, continuous learning and improvement are essential for maintaining the safety of AI systems. AI systems should be designed to learn from their experiences and updates continuously, incorporating new data to enhance their accuracy and reliability. This ongoing learning process helps AI systems adapt to changing environments and emerging threats, ensuring sustained safety and performance. In healthcare AI, continuous learning allows systems to update their diagnostic algorithms based on new medical research and clinical data, improving patient outcomes and safety.

In conclusion, the safety of AI systems is paramount and requires a multifaceted approach that includes redundancy, real-time monitoring, fail-safe mechanisms, transparency, ethical programming, cybersecurity, and continuous learning. By integrating these safety features, AI systems can operate reliably and securely, fostering trust and ensuring that they enhance, rather than endanger, human lives. The commitment to safety in AI development is crucial for realizing the full potential of these technologies in a responsible and ethical manner.

Monitoring and Compliance

Monitoring and compliance are crucial elements in the deployment and operation of artificial intelligence systems. Ensuring that AI systems adhere to established ethical guidelines and performance standards is essential for maintaining trust, safety, and efficacy. Continuous monitoring and stringent compliance measures help prevent misuse, detect anomalies, and address issues promptly, ensuring that AI systems operate within safe and ethical boundaries.

The process of monitoring AI systems involves the continuous observation of their performance and behavior. This is achieved through various tools and methodologies that track the system's actions, decisions, and outcomes. Monitoring helps to identify any deviations from expected behavior, which can be indicators of potential problems such as bias, inaccuracies, or technical malfunctions. For instance, in autonomous vehicles, real-time monitoring systems can track the vehicle's movements, sensor inputs, and decision-making processes to ensure it operates safely and responds appropriately to dynamic road conditions.

Compliance, on the other hand, involves ensuring that AI systems adhere to regulatory standards, ethical guidelines, and best practices. This includes compliance with data privacy laws, such as the General Data Protection Regulation (GDPR), which governs the collection, storage, and use of personal data. Compliance also extends to industry-specific regulations, such as those in healthcare, finance, or transportation, which set specific standards for safety, accuracy, and accountability.

One of the key challenges in monitoring and compliance is ensuring transparency and explainability in AI systems. AI, particularly those based on complex machine learning algorithms, often operates as a "black box," making it difficult to understand how decisions are made. Enhancing transparency involves developing methods to interpret and explain AI decision-making processes. This can include the use of interpretable models, providing clear documentation of the algorithms, and implementing tools that visualize how inputs are processed to produce outputs. Transparency is crucial for both monitoring and compliance, as it allows stakeholders to assess whether the AI system is functioning correctly and ethically.

Another important aspect of monitoring and compliance is the establishment of clear accountability frameworks. This involves defining roles and responsibilities for individuals and organizations involved in the development, deployment, and oversight of AI systems. Accountability frameworks ensure that there is a clear understanding of who is responsible for various aspects of the AI system's operation, from data management to decision-making to handling errors and malfunctions. In the event of an issue, these frameworks facilitate swift identification of the responsible parties and the implementation of corrective measures.

To support effective monitoring and compliance, organizations often establish internal audit and review processes. These processes involve regular evaluations of the AI system's performance, adherence to standards, and impact on users and society. Audits can be conducted by internal teams or external third parties to ensure objectivity and thoroughness. The findings from these audits are used to refine the AI system, address any identified issues, and improve overall compliance with ethical and regulatory requirements.

In addition to internal measures, regulatory bodies play a critical role in overseeing AI systems. Governments and international organizations are increasingly recognizing the need for robust regulatory frameworks to govern the use of AI. These frameworks set out the legal and ethical standards that AI systems must meet and establish mechanisms for enforcement and accountability. Regulatory oversight can include mandatory reporting of AI system performance, regular inspections, and the imposition of penalties for non-compliance. By creating a strong regulatory environment, authorities can ensure that AI systems are developed and deployed in a manner that prioritizes public safety and ethical considerations.

Moreover, the integration of feedback mechanisms into AI systems is essential for continuous improvement. Feedback from users, stakeholders, and monitoring systems provides valuable insights into the AI system's performance and impact. This feedback can highlight areas where the system excels and identify aspects that require improvement. Incorporating feedback into the development cycle ensures that AI systems evolve to meet changing needs and expectations while adhering to ethical and regulatory standards.

In conclusion, effective monitoring and compliance are foundational to the ethical and safe deployment of AI systems. Through continuous observation, adherence to regulatory standards, transparency, accountability frameworks, and feedback mechanisms, organizations can ensure that their AI systems operate within defined ethical boundaries and deliver reliable, trustworthy outcomes. This comprehensive approach to monitoring and compliance not only mitigates risks but also fosters public trust and supports the responsible advancement of AI technology.

Feedback and Improvement Mechanisms

Artificial intelligence systems are continuously evolving, and one of the most critical aspects of their development is the integration of effective feedback and improvement mechanisms. These mechanisms ensure that AI systems remain reliable, efficient, and aligned with ethical standards as they adapt to new data and environments.

Feedback loops are essential for the dynamic adjustment and enhancement of AI systems. These loops function by collecting data on the system's performance, user interactions, and environmental changes. This data is then analyzed to identify patterns, trends, and areas for improvement. For instance, in natural language processing systems, user feedback on the accuracy and relevance of responses can be used to refine algorithms, improving the system's ability to understand and generate human language.

The implementation of robust feedback mechanisms involves multiple stages, starting with data collection. This stage requires the systematic gathering of diverse data points that reflect the AI system's performance. This data can come from various sources, including user interactions, sensor readings, and system logs. Ensuring the quality and representativeness of this data is crucial, as it forms the foundation for subsequent analysis and improvements.

Once the data is collected, the next step is analysis. Advanced analytical tools and techniques, such as machine learning algorithms, are used to process and interpret the data. These tools can identify inefficiencies, errors, and potential biases in the AI system. For example, in autonomous vehicles, analysis of driving data can reveal situations where the system's performance deviates from expected behavior, indicating areas that require adjustment or further training.

The insights gained from data analysis are then used to inform the improvement of the AI system. This involves updating algorithms, adjusting parameters, and enhancing system capabilities. Continuous learning algorithms play a vital role in this process, enabling AI systems to automatically incorporate new data and improve over time. This iterative process ensures that AI systems remain up-to-date and capable of handling emerging challenges and opportunities.

User feedback is another critical component of improvement mechanisms. Engaging users in the feedback process provides valuable insights into the AI system's practical performance and usability. For instance, users of a healthcare AI system can provide feedback on the system's diagnostic accuracy and ease of use, which can then be used to refine its algorithms and interface. Creating channels for users to easily report issues and suggest improvements is essential for harnessing this valuable input.

Transparency and accountability are fundamental to effective feedback and improvement mechanisms. AI developers must ensure that users understand how their feedback is used and the impact it has on the system's evolution. This transparency builds trust and encourages continued user engagement. Additionally, establishing accountability frameworks ensures that feedback is systematically addressed and that improvements are implemented in a timely and effective manner.

Ethical considerations are also paramount in the feedback and improvement process. Ensuring that AI systems respect user privacy and data security is crucial when collecting and analyzing feedback. Implementing robust data anonymization and protection measures helps safeguard user information while enabling valuable insights to be drawn from the data.

Moreover, addressing biases in AI systems is a key ethical concern. Feedback mechanisms must include checks for biases that may be present in the data or the system's outputs. Regular audits and updates based on feedback can help mitigate biases, ensuring that AI systems operate fairly and equitably. For example, in hiring algorithms, feedback on hiring decisions can reveal potential biases, leading to adjustments that promote diversity and fairness.

In conclusion, feedback and improvement mechanisms are essential for the sustained success and ethical deployment of AI systems. By systematically collecting and analyzing data, engaging users in the feedback process, and ensuring transparency and accountability, AI systems can continuously evolve and improve. These mechanisms help maintain the relevance, reliability, and ethical integrity of AI systems, ensuring that they meet the needs and expectations of users and society at large.

Chapter 2: The Supremacy of the First Directive

Prioritizing Human Safety

Conflict Resolution in AI

In the realm of artificial intelligence, conflict resolution stands as a critical challenge, necessitating sophisticated mechanisms to ensure systems operate harmoniously with human values and societal norms. AI systems, given their complexity and the diverse environments they navigate, often encounter conflicting objectives and ethical dilemmas. Resolving these conflicts effectively is paramount to maintaining trust, safety, and reliability in AI applications.

One of the primary methods for addressing conflict in AI is through the design of robust ethical frameworks that guide decision-making processes. These frameworks incorporate principles that prioritize human welfare, fairness, and transparency. For instance, in autonomous vehicles, ethical decision-making frameworks are essential to resolve conflicts that arise in split-second scenarios, such as choosing between different evasive maneuvers to avoid an accident. These frameworks ensure that the vehicle's responses are aligned with broader ethical standards, minimizing harm and prioritizing human safety.

Another critical aspect of conflict resolution in AI involves the implementation of advanced algorithms that can balance competing objectives. Multi-objective optimization techniques allow AI systems to evaluate various factors and make decisions that strike a balance between conflicting goals. For example, in resource allocation for disaster response, AI systems must balance the urgency of different needs, the availability of resources, and logistical constraints to optimize the distribution of aid. These algorithms are designed to find the most equitable and effective solutions, ensuring that resources are utilized efficiently and ethically.

Transparency and explainability are also crucial in resolving conflicts within AI systems. By making the decision-making processes of AI transparent and understandable, developers can ensure that users and stakeholders trust and accept the system's actions. Explainable AI techniques involve creating models that provide clear, interpretable explanations for their decisions. This transparency helps users understand how conflicts are resolved and allows them to challenge or question decisions that appear unjust or erroneous. For instance, in AI-driven judicial systems, providing clear explanations for sentencing recommendations ensures that the judicial process remains fair and accountable.

Human oversight remains an indispensable component of conflict resolution in AI. While AI systems can process vast amounts of data and make complex decisions, human judgment is essential in overseeing these decisions, especially in ethically charged situations. Human-in-the-loop approaches involve incorporating human judgment at critical points in the decision-making process, allowing for the resolution of conflicts that may be beyond the scope of the AI's capabilities. This collaborative approach ensures that AI systems operate within acceptable ethical boundaries and that their actions align with human values.

In addition to these technical and procedural measures, continuous monitoring and feedback mechanisms are vital for effective conflict resolution in AI. Regular monitoring allows for the detection of emerging conflicts and issues that may not have been anticipated during the system's development. Feedback loops enable AI systems to learn from past conflicts and adjust their behavior accordingly. For example, in customer service chatbots, continuous monitoring of interactions and feedback from users help refine the system's responses, ensuring that conflicts are resolved promptly and satisfactorily.

Moreover, establishing clear regulatory frameworks and standards is essential for guiding AI systems in conflict resolution. Governments and regulatory bodies play a crucial role in defining the ethical and legal boundaries within which AI systems must operate. These regulations ensure that AI systems are designed and deployed in ways that respect human rights, promote fairness, and prevent harm. Compliance with these standards provides a structured approach to conflict resolution, ensuring that AI systems contribute positively to society.

Ethical training and awareness among AI developers and operators are also important for effective conflict resolution. Ensuring that those involved in the development and deployment of AI systems are aware of ethical considerations and trained in ethical decision-making helps prevent conflicts from arising and equips them to handle conflicts appropriately when they do occur. This human element reinforces the ethical integrity of AI systems and ensures that they are aligned with societal values.

In conclusion, conflict resolution in AI involves a multifaceted approach that combines robust ethical frameworks, advanced algorithms, transparency, human oversight, continuous monitoring, regulatory compliance, and ethical training. By integrating these elements, AI systems can navigate complex conflicts effectively, maintaining trust and ensuring that their operations are aligned with human values and societal norms. This comprehensive approach is essential for the responsible and ethical advancement of AI technologies.

Regulatory Standards

The development and deployment of artificial intelligence systems have advanced rapidly, bringing about significant transformations across various sectors. However, the integration of these technologies necessitates the establishment of robust regulatory standards to ensure their ethical use, safety, and compliance with societal norms. Regulatory standards in AI play a crucial role in safeguarding public interests, promoting fairness, and mitigating risks associated with AI applications.

One of the primary objectives of regulatory standards in AI is to protect privacy and data security. AI systems often rely on vast amounts of data, some of which may be sensitive or personal.

Regulations such as the General Data Protection Regulation (GDPR) in Europe set stringent requirements for data protection, ensuring that individuals' privacy is maintained and that their data is processed transparently and securely. These regulations mandate that organizations obtain explicit consent from individuals before collecting their data, anonymize data where possible, and implement robust security measures to prevent data breaches.

Transparency and accountability are also central tenets of regulatory standards for AI. Transparency involves making AI systems' decision-making processes understandable to users and stakeholders. Regulations often require that AI developers provide clear documentation of how their systems work, the data they use, and the algorithms they employ. This transparency helps build trust and allows for the auditing of AI systems to ensure they operate as intended. Accountability, on the other hand, ensures that there are clear lines of responsibility for the actions and outcomes of AI systems. This includes establishing mechanisms for redress if an AI system causes harm or operates unfairly.

Bias and discrimination are critical issues that regulatory standards aim to address. AI systems can inadvertently perpetuate and amplify existing biases present in their training data, leading to unfair or discriminatory outcomes. Regulatory frameworks mandate the regular testing and auditing of AI systems for biases, ensuring that they are identified and mitigated. This involves developing diverse and representative training datasets and implementing fairness metrics to evaluate the performance of AI systems across different demographic groups. By addressing bias, regulations help ensure that AI systems operate equitably and do not reinforce societal inequalities.

Safety and reliability are paramount concerns in the deployment of AI, particularly in high-stakes environments such as healthcare, transportation, and finance. Regulatory standards establish safety protocols that AI systems must adhere to, including rigorous testing and validation processes before deployment. For instance, autonomous vehicles must undergo extensive testing in various driving conditions to ensure they can operate safely on public roads. Similarly, AI systems used in healthcare must be validated for accuracy and reliability to prevent misdiagnoses and ensure patient safety.

Ethical considerations are integral to the regulatory landscape for AI. Standards often incorporate ethical guidelines that AI systems must follow, ensuring they respect human rights and operate in alignment with societal values. This includes principles such as beneficence, non-maleficence, and justice. Ethical AI frameworks guide developers in designing systems that prioritize human well-being, avoid causing harm, and distribute benefits and risks fairly. By embedding ethical considerations into regulatory standards, societies can ensure that AI technologies contribute positively to human flourishing.

Interoperability and standardization are also important aspects of AI regulation. Interoperability ensures that different AI systems can work together seamlessly, facilitating innovation and collaboration across sectors. Standardization involves developing common protocols and benchmarks for AI systems, ensuring consistency and reliability in their performance. Regulatory bodies often work with industry stakeholders to develop these standards, fostering an ecosystem where AI technologies can evolve harmoniously and effectively.

Continuous monitoring and evaluation are essential components of regulatory frameworks for AI. Regulations require ongoing oversight of AI systems to ensure they continue to comply with standards and adapt to emerging challenges. This involves regular audits, performance reviews, and updates to regulatory guidelines to keep pace with technological advancements. Continuous monitoring helps identify and address issues proactively, ensuring that AI systems remain safe, reliable, and ethical over time.

International cooperation and harmonization of regulatory standards are crucial for the global governance of AI. As AI technologies are developed and deployed across borders, consistent regulatory standards help prevent regulatory arbitrage and ensure a level playing field. International organizations, such as the United Nations and the International Telecommunication Union, work towards developing global frameworks for AI governance, promoting cooperation and consistency in regulatory approaches.

In conclusion, regulatory standards are essential for the responsible development and deployment of AI systems. By ensuring privacy and data security, promoting transparency and accountability, addressing bias and discrimination, safeguarding safety and reliability, embedding ethical considerations, and fostering interoperability and standardization, regulatory frameworks help mitigate risks and enhance the benefits of AI technologies. Continuous monitoring and international cooperation further strengthen these efforts, ensuring that AI serves humanity in the most ethical and beneficial ways possible. Through robust regulatory standards, societies can harness the transformative potential of AI while protecting public interests and upholding fundamental values.

Ethical Dilemmas

Artificial intelligence presents a myriad of ethical dilemmas that challenge developers, users, and policymakers alike. As AI systems become more sophisticated and integrated into society, the complexity of these dilemmas intensifies. Addressing ethical dilemmas requires a nuanced understanding of the underlying principles and the potential consequences of AI deployment.

One of the most prominent ethical dilemmas in AI is the balance between privacy and utility. AI systems, particularly those powered by machine learning, rely on large datasets to function effectively. These datasets often contain personal and sensitive information, raising concerns about privacy and data security. For example, AI algorithms used in healthcare can significantly improve diagnostics and treatment plans by analyzing patient data. However, the collection and analysis of such data can infringe on patient privacy if not handled with stringent safeguards. The ethical challenge lies in maximizing the benefits of AI while ensuring robust protections for individual privacy.

Another ethical dilemma involves the potential for bias and discrimination in AI systems. AI algorithms learn from historical data, which can reflect existing societal biases. If these biases are not identified and mitigated, AI systems can perpetuate and even amplify discrimination. This issue is particularly critical in areas such as hiring, lending, and law enforcement, where biased algorithms can lead to unfair treatment of certain groups. Ensuring fairness and equity in AI requires careful consideration of the training data, continuous monitoring for biased outcomes, and the development of techniques to counteract biases.

The question of accountability in AI decision-making also presents a significant ethical dilemma. As AI systems become more autonomous, determining responsibility for their actions becomes increasingly complex. In scenarios where AI systems make critical decisions, such as in autonomous driving or medical diagnostics, assigning accountability for errors or harmful outcomes is challenging. Establishing clear frameworks for accountability is essential to ensure that there are mechanisms for redress and that ethical standards are upheld. This may involve defining the roles and responsibilities of AI developers, users, and regulatory bodies.

Transparency is another key ethical consideration in AI. The opacity of many AI systems, particularly those using deep learning techniques, makes it difficult for users to understand how decisions are made. This lack of transparency can undermine trust and make it challenging to identify and correct errors. Ensuring transparency involves developing explainable AI models that provide clear and understandable justifications for their decisions. This is crucial in high-stakes applications, such as healthcare and criminal justice, where the consequences of AI decisions can be profound.

The deployment of AI in warfare and autonomous weapons systems raises profound ethical questions about the nature of conflict and the value of human life. Autonomous weapons, capable of making decisions without human intervention, challenge traditional notions of accountability and the laws of war.

The ethical dilemma centers on whether it is morally acceptable to delegate life-and-death decisions to machines. International discussions and treaties are essential to address these issues, ensuring that the use of AI in warfare adheres to humanitarian principles and international law.

Another ethical dilemma is the impact of AI on employment and the workforce. While AI has the potential to enhance productivity and create new job opportunities, it also poses the risk of displacing workers through automation. The ethical challenge is to balance the economic benefits of AI with the need to support workers whose jobs are affected. This involves rethinking education and training programs to equip the workforce with the skills needed in an AI-driven economy and implementing policies that support workers during transitions.

The use of AI in surveillance and the potential for intrusive monitoring also present ethical challenges. AI-powered surveillance systems can enhance security and aid in crime prevention, but they also pose risks to civil liberties and the right to privacy. The ethical dilemma is how to balance security needs with the protection of individual freedoms. This requires establishing legal frameworks that regulate the use of surveillance technologies, ensuring they are used proportionately and with appropriate safeguards.

In conclusion, the ethical dilemmas posed by AI are complex and multifaceted, requiring a comprehensive and nuanced approach. Addressing these dilemmas involves balancing competing interests, ensuring fairness and accountability, promoting transparency, and protecting human rights. As AI continues to evolve, ongoing dialogue and collaboration among stakeholders, including technologists, ethicists, policymakers, and the public, are essential to navigate these ethical challenges and ensure that AI is developed and deployed in a manner that benefits society while upholding ethical principles.

Balancing Conflicting Objectives

Multi-objective Optimization

In the evolving landscape of artificial intelligence, multi-objective optimization plays a crucial role in ensuring that AI systems can effectively balance competing objectives. This approach allows AI to navigate complex decision-making scenarios where various goals must be considered simultaneously, often involving trade-offs between conflicting interests. Understanding and implementing multi-objective optimization is essential for developing AI that is both efficient and ethically sound.

Multi-objective optimization involves the simultaneous optimization of two or more conflicting objectives. Unlike single-objective optimization, which focuses on optimizing one goal, multi-objective optimization requires finding a balance between multiple goals, often without a clear priority. This process is critical in many real-world applications where decisions impact various stakeholders and aspects of the system.

One prominent example of multi-objective optimization is in the field of autonomous vehicles. These vehicles must balance safety, efficiency, passenger comfort, and adherence to traffic laws. For instance, an autonomous car must navigate in a way that minimizes travel time while also ensuring the safety of passengers and pedestrians. In scenarios where these objectives conflict, such as deciding whether to swerve to avoid an obstacle, the vehicle must optimize for the best overall outcome.

The process of multi-objective optimization in AI typically involves several steps. The first step is defining the objectives clearly. These objectives must be quantifiable and relevant to the problem at hand. In the context of autonomous vehicles, objectives might include minimizing travel time, maximizing safety, and reducing fuel consumption. Each objective is assigned a function that the AI system will optimize.

The next step is to establish constraints. Constraints are conditions that the AI system must adhere to while optimizing its objectives. For autonomous vehicles, constraints might include speed limits, legal regulations, and the physical limitations of the vehicle. These constraints ensure that the AI system operates within acceptable boundaries and does not compromise essential safety and ethical standards.

Once the objectives and constraints are defined, the AI system employs optimization algorithms to find the best possible solutions. These algorithms explore the solution space to identify trade-offs between conflicting objectives. Common algorithms used in multi-objective optimization include genetic algorithms, particle swarm optimization, and evolutionary strategies. These algorithms are designed to handle the complexity and non-linearity of real-world problems, providing robust solutions that balance multiple objectives effectively.

An essential aspect of multi-objective optimization is the concept of Pareto efficiency. A solution is considered Pareto efficient if no objective can be improved without worsening another objective. The set of all Pareto-efficient solutions forms the Pareto frontier, which represents the best possible trade-offs between objectives. Decision-makers can then select a solution from the Pareto frontier based on their preferences and priorities.

In addition to technical considerations, ethical implications play a significant role in multi-objective optimization. AI systems must be designed to ensure that optimization does not lead to unfair or harmful outcomes. For example, in healthcare, AI systems used for resource allocation must balance efficiency with fairness, ensuring that all patients receive appropriate care. Ethical frameworks guide the optimization process, helping to align AI decisions with societal values and norms.

Continuous learning and adaptation are also crucial for effective multi-objective optimization. AI systems must be capable of learning from new data and experiences, updating their optimization strategies accordingly. This adaptability ensures that the AI system remains effective in dynamic environments where objectives and constraints may change over time. For instance, an AI system managing a smart city must continuously adapt to evolving traffic patterns, weather conditions, and regulatory changes to optimize urban mobility and sustainability.

Transparency and explainability are essential components of multi-objective optimization. Stakeholders must understand how AI systems make decisions and balance competing objectives. Transparent algorithms and clear documentation help build trust and ensure accountability. For example, in financial AI systems that optimize investment portfolios, transparency about the trade-offs and risks involved in different investment strategies is crucial for gaining the confidence of investors.

In conclusion, multi-objective optimization is a fundamental aspect of AI development that enables systems to balance competing goals effectively. By clearly defining objectives and constraints, employing robust optimization algorithms, ensuring ethical considerations, and maintaining transparency and adaptability, AI systems can navigate complex decision-making scenarios and deliver balanced, efficient, and fair outcomes. As AI continues to advance, multi-objective optimization will remain a critical tool for addressing the diverse and dynamic challenges of real-world applications.

Real-world Scenarios

Artificial intelligence systems have become integral to various aspects of modern life, from healthcare to finance to autonomous transportation. To understand how these technologies function effectively in the real world, it is essential to examine real-world scenarios where AI systems are deployed and how they address complex challenges.

One of the most significant real-world applications of AI is in healthcare, where AI systems assist in diagnosing diseases, personalizing treatment plans, and managing patient care. For instance, AI algorithms can analyze medical images to detect anomalies such as tumors with remarkable accuracy. In a real-world scenario, a radiologist might use an AI-powered diagnostic tool to identify early signs of cancer in mammograms. The AI system processes thousands of images, learning to distinguish between benign and malignant growths. This assistance not only improves diagnostic accuracy but also accelerates the process, allowing for earlier intervention and better patient outcomes. However, these systems must be rigorously tested to ensure they do not produce false positives or negatives, which could have serious implications for patient care.

In the realm of autonomous vehicles, AI is revolutionizing transportation by enabling cars to navigate and make decisions without human intervention. Autonomous vehicles rely on a combination of sensors, cameras, and machine learning algorithms to understand their environment and respond appropriately. A real-world scenario might involve an autonomous car navigating through a busy urban area. The AI system must recognize traffic signals, pedestrians, cyclists, and other vehicles, making real-time decisions to ensure safety and efficiency. For instance, if a pedestrian suddenly steps into the street, the AI must quickly calculate the safest way to stop or maneuver to avoid an accident. This requires the AI to balance multiple objectives such as passenger safety, traffic laws, and efficiency, showcasing the complexity of real-world AI applications.

AI is also making significant strides in financial services, particularly in fraud detection and credit scoring. Financial institutions deploy AI systems to monitor transactions for suspicious activities that may indicate fraud. In a real-world scenario, an AI system might analyze patterns in credit card transactions to detect anomalies that suggest fraudulent behavior. The system must be able to distinguish between legitimate transactions and fraudulent ones, even as fraudsters constantly evolve their tactics. This involves real-time data processing and the ability to adapt to new fraud patterns, demonstrating the AI's capability to learn and improve over time.

In environmental management, AI technologies are being used to monitor and address climate change and resource conservation. For example, AI systems can analyze satellite imagery to track deforestation, predict weather patterns, and manage natural resources. In a real-world scenario, an AI system might be used to monitor a forest for illegal logging activities. By analyzing data from satellite images, the AI can identify areas of deforestation and alert authorities to take action. This application not only helps in conserving biodiversity but also aids in enforcing environmental laws and regulations.

Another critical application of AI is in disaster response and management. AI systems can predict natural disasters, coordinate relief efforts, and optimize resource allocation. For instance, an AI system might be used to predict the path of a hurricane and its potential impact on coastal communities. This information can be used to coordinate evacuation plans, deploy emergency services, and allocate resources effectively. In a real-world scenario, AI can analyze weather data, historical storm patterns, and geographical information to provide accurate predictions, enabling timely and informed decision-making during disasters.

In the field of agriculture, AI is enhancing productivity and sustainability through precision farming techniques. AI systems analyze data from various sources, including soil sensors, weather forecasts, and crop health images, to optimize farming practices. A real-world scenario might involve an AI system advising farmers on the optimal time to plant or harvest crops based on soil moisture levels and weather predictions. This not only maximizes yield but also conserves resources such as water and fertilizers, promoting sustainable agricultural practices.

These real-world scenarios highlight the transformative potential of AI across different sectors. However, the deployment of AI systems in these scenarios also underscores the importance of addressing ethical and technical challenges. Ensuring the accuracy, fairness, and transparency of AI systems is crucial for their effective and ethical application. Continuous monitoring, rigorous testing, and adherence to regulatory standards are essential to mitigate risks and enhance the benefits of AI technologies.

In conclusion, AI systems are increasingly being integrated into real-world applications, offering significant benefits in various domains. From healthcare to autonomous vehicles to environmental management, AI is helping to solve complex problems and improve efficiency. However, the successful deployment of AI in these scenarios requires careful consideration of ethical implications, robust testing, and adherence to regulatory frameworks. By addressing these challenges, AI can continue to advance and contribute positively to society.

Algorithmic Transparency

In the development and deployment of artificial intelligence systems, algorithmic transparency is a critical component that fosters trust, accountability, and ethical use. As AI technologies become more embedded in various aspects of society, from healthcare to finance to criminal justice, the importance of understanding how these algorithms work cannot be overstated. Algorithmic transparency involves making the inner workings of AI systems clear and understandable to users, stakeholders, and regulators.

One of the main reasons algorithmic transparency is essential is to build trust between AI systems and their users. When people understand how an AI system reaches its decisions, they are more likely to trust and accept its outcomes. For instance, in the context of healthcare, an AI system that diagnoses medical conditions must be able to explain its reasoning to doctors and patients. This transparency not only helps in validating the AI's decisions but also enables healthcare professionals to trust the system and integrate it into their practice.

Transparency also plays a vital role in accountability. When AI systems are used to make significant decisions, such as approving loans, hiring employees, or determining parole eligibility, it is crucial that these decisions can be audited and reviewed. Algorithmic transparency ensures that stakeholders can trace the decision-making process, identify any errors or biases, and hold the developers accountable. This level of accountability is particularly important in preventing and addressing any unfair or discriminatory practices that may arise from the use of AI.

The challenge of algorithmic transparency is often linked to the complexity of AI models, especially those based on deep learning. These models, while highly effective, can operate as "black boxes," where their decision-making processes are not easily interpretable. Addressing this challenge requires the development of techniques and tools that can provide insights into how these models function. Methods such as model interpretability and explainability techniques have been developed to shed light on the internal workings of complex AI systems.

Model interpretability involves simplifying the AI model or using surrogate models that approximate the behavior of the original model in a more understandable way. For example, decision trees and linear models are often used as interpretable models because their decision paths can be easily visualized and understood. These surrogate models can help explain the decisions of more complex models by showing similar decision patterns in a simpler format.

Explainability techniques, on the other hand, focus on providing post-hoc explanations for specific decisions made by AI systems. Techniques like LIME (Local Interpretable Model-agnostic Explanations) and SHAP (SHapley Additive exPlanations) are used to analyze individual predictions and explain which features most influenced the outcome. For instance, if an AI system denies a loan application, these techniques can identify the key factors that led to this decision, such as credit score, income level, or employment history, providing a clear and understandable rationale.

Ethical considerations are at the forefront of algorithmic transparency. Ensuring that AI systems operate transparently aligns with ethical principles of fairness, accountability, and respect for human rights. Transparency helps to uncover and mitigate biases in AI systems, ensuring that they do not perpetuate or exacerbate existing inequalities. For example, in predictive policing, transparency can help ensure that AI systems do not disproportionately target certain communities, promoting fairness and equity in law enforcement.

Regulatory frameworks increasingly emphasize the need for algorithmic transparency. Regulations like the GDPR mandate that individuals have the right to obtain meaningful information about the logic involved in automated decisions that affect them. Compliance with such regulations requires organizations to implement transparent AI practices and provide clear explanations for their automated decision-making processes. This regulatory pressure ensures that transparency is not just a best practice but a legal requirement for ethical AI deployment.

Moreover, transparency fosters innovation and collaboration. When AI models and their decision-making processes are open and understandable, researchers and developers can build on existing work, identify areas for improvement, and collaborate to enhance AI systems. This openness accelerates technological advancement and ensures that innovations are shared and built upon within the AI community.

In conclusion, algorithmic transparency is a fundamental aspect of ethical and effective AI deployment. By making AI systems understandable and accountable, transparency fosters trust, ensures fairness, and aligns with regulatory and ethical standards. Through techniques like model interpretability and explainability, the AI community can address the challenges of complex models and promote a culture of openness and accountability. As AI continues to advance, maintaining a commitment to transparency will be crucial in ensuring that these technologies benefit society while upholding ethical principles and human rights.

Policy and Governance

Role of Government

The role of government in the development and regulation of artificial intelligence is multifaceted and critical to ensuring that AI technologies are used ethically and effectively. Governments have a unique responsibility to balance the promotion of innovation with the protection of public interests, including privacy, security, and equity. As AI continues to permeate various sectors, from healthcare to finance to national security, the importance of robust government oversight and regulation cannot be overstated.

One of the primary roles of government in AI is to establish and enforce regulatory frameworks that ensure the safe and ethical use of these technologies. This involves creating laws and guidelines that govern the development, deployment, and operation of AI systems. For instance, regulations may mandate rigorous testing and validation processes for AI applications, especially those used in high-stakes environments such as autonomous vehicles or medical diagnostics. By setting these standards, governments help prevent the deployment of faulty or biased AI systems that could harm individuals or society at large.

Governments also play a crucial role in protecting data privacy and security, which are paramount concerns in the age of AI. With AI systems often relying on vast amounts of personal data to function effectively, there is a significant risk of data breaches and misuse. Governments can address these risks by enacting comprehensive data protection laws, such as the General Data Protection Regulation (GDPR) in the European Union. These laws set stringent requirements for how data is collected, stored, and used, ensuring that individuals' privacy is safeguarded. Additionally, governments can establish oversight bodies to monitor compliance with these regulations and take action against entities that violate them.

Another important role of government is to promote transparency and accountability in AI systems. Transparency involves making the decision-making processes of AI systems clear and understandable to users and stakeholders. This is particularly important in sectors where AI decisions can have significant impacts, such as criminal justice or finance. Governments can require that organizations using AI provide explanations for their systems' decisions, enabling individuals to understand how and why decisions were made. This transparency not only builds trust but also allows for the identification and correction of errors or biases in AI systems.

Accountability is closely linked to transparency and involves ensuring that there are mechanisms in place to hold AI developers and users responsible for the outcomes of their systems. Governments can establish legal frameworks that define liability for AI-related harms, ensuring that affected individuals have avenues for redress. This may involve updating existing laws to address the unique challenges posed by AI or creating new legal categories specifically for AI-related issues. By holding entities accountable, governments help ensure that AI systems are developed and used in a manner that prioritizes public welfare.

Governments also have a vital role in fostering innovation and ensuring that the benefits of AI are widely distributed. This involves investing in research and development, supporting education and training programs, and providing incentives for companies to innovate responsibly. Governments can fund AI research initiatives that address societal challenges, such as healthcare, climate change, and public safety. By supporting education and training programs, governments help build a skilled workforce capable of developing and managing AI technologies. Additionally, providing incentives for companies to adopt ethical AI practices encourages the private sector to align with public interests.

Moreover, governments can play a proactive role in addressing the societal impacts of AI, such as job displacement and inequality. AI has the potential to automate many tasks, leading to significant changes in the labor market. Governments can implement policies to support workers affected by automation, such as retraining programs and social safety nets. By proactively addressing these impacts, governments help ensure that the transition to an AI-driven economy is inclusive and equitable.

International cooperation is another critical aspect of the government's role in AI. AI technologies often cross national borders, and their impacts are felt globally. Governments must work together to establish international standards and frameworks for AI governance. This cooperation can take the form of international treaties, agreements, or collaborative research initiatives. By working together, governments can address global challenges, such as cybersecurity and ethical AI deployment, more effectively.

In conclusion, the role of government in AI is crucial to ensuring that these technologies are developed and used in ways that benefit society while minimizing risks. Through regulatory frameworks, data protection laws, transparency and accountability measures, support for innovation, and international cooperation, governments can guide the responsible development and deployment of AI. As AI continues to evolve, the government's role will be essential in balancing the opportunities and challenges presented by this transformative technology.

Industry Standards

Artificial intelligence and robotics have advanced significantly over the past decades, and with these advancements, the need for robust industry standards has become increasingly evident. Industry standards in AI ensure that technologies are developed and deployed in a manner that prioritizes safety, reliability, and ethical considerations. These standards help create a cohesive framework that guides AI developers, users, and regulators, fostering innovation while protecting public interests.

Industry standards serve as benchmarks for quality and performance, ensuring that AI systems meet specific criteria before they are widely adopted. These standards cover various aspects of AI development, including data management, algorithm design, system testing, and user interaction. By adhering to these standards, companies can produce AI technologies that are reliable, safe, and effective. For example, in the automotive industry, standards for autonomous vehicles ensure that these systems can operate safely under diverse conditions, reducing the risk of accidents and enhancing public trust.

One of the critical components of industry standards is data quality and integrity. AI systems rely heavily on data to function effectively, and the quality of this data directly impacts the system's performance. Standards for data management include guidelines for data collection, storage, processing, and usage. These guidelines ensure that the data used to train AI models is accurate, complete, and representative of real-world scenarios. For instance, in healthcare, standards for medical data ensure that patient information is accurately recorded and securely stored, enabling AI systems to provide precise and reliable diagnostics and treatment recommendations.

Algorithmic transparency and fairness are also essential aspects of industry standards. As AI systems become more complex, it is crucial to ensure that their decision-making processes are transparent and unbiased. Standards in this area require developers to implement measures that make AI algorithms understandable and auditable. This includes providing documentation that explains how algorithms work, the data they use, and the logic behind their decisions. Additionally, standards for fairness require that AI systems are tested for biases and that steps are taken to mitigate any identified biases. For example, in hiring algorithms, standards ensure that the systems do not unfairly discriminate against candidates based on race, gender, or other protected characteristics.

Interoperability is another vital consideration in industry standards. As AI systems are integrated into various sectors, ensuring that these systems can work seamlessly with existing technologies is crucial. Standards for interoperability provide guidelines for the development of AI systems that can communicate and interact with other systems, promoting a cohesive and efficient technological ecosystem. In the context of smart cities, interoperability standards ensure that AI systems used for traffic management, energy distribution, and public safety can work together harmoniously, enhancing the overall functionality and efficiency of urban infrastructure.

Cybersecurity standards are critical in protecting AI systems from malicious attacks and ensuring the integrity of their operations. As AI systems become more interconnected and reliant on data, they become potential targets for cyber threats. Industry standards for cybersecurity include guidelines for securing data, protecting against unauthorized access, and ensuring the resilience of AI systems against attacks. These standards help safeguard sensitive information and maintain the reliability and trustworthiness of AI technologies. For example, in financial services, cybersecurity standards ensure that AI systems used for transaction monitoring and fraud detection are secure and resilient against hacking attempts.

Ethical considerations are at the heart of industry standards for AI. These standards ensure that AI systems are developed and used in ways that align with societal values and ethical principles. This includes guidelines for respecting user privacy, ensuring fairness and equity, and preventing harm. Ethical standards help build public trust in AI technologies and ensure that their deployment contributes positively to society. For instance, in facial recognition technology, ethical standards ensure that the systems are used responsibly, with respect for privacy and without infringing on individual rights.

International cooperation is essential in establishing and maintaining industry standards for AI. As AI technologies are developed and deployed globally, harmonizing standards across countries ensures consistency and facilitates international collaboration. Organizations such as the International Organization for Standardization (ISO) and the Institute of Electrical and Electronics Engineers (IEEE) play a crucial role in developing global standards for AI. These organizations bring together experts from various fields to create comprehensive and widely accepted guidelines that govern the development and use of AI technologies.

In conclusion, industry standards are fundamental to the responsible development and deployment of AI. By providing guidelines for data quality, algorithmic transparency, interoperability, cybersecurity, and ethical considerations, these standards ensure that AI technologies are reliable, safe, and aligned with societal values. International cooperation in establishing and maintaining these standards further enhances their effectiveness, promoting a cohesive and collaborative approach to AI development. As AI continues to evolve, adherence to robust industry standards will be crucial in harnessing its potential for the benefit of society.

Global Cooperation

Global cooperation is fundamental to the ethical development and deployment of artificial intelligence. As AI technologies increasingly impact every aspect of our lives, from healthcare and finance to national security and beyond, it becomes imperative for countries to collaborate and create unified standards and frameworks. This collaboration ensures that AI is used responsibly, ethically, and for the global good, addressing challenges that transcend national boundaries.

One of the primary areas where global cooperation is crucial is in the establishment of international standards for AI development and deployment. Organizations such as the International Organization for Standardization (ISO) and the Institute of Electrical and Electronics Engineers (IEEE) play a significant role in developing these standards. These bodies bring together experts from around the world to create comprehensive guidelines that ensure AI systems are safe, reliable, and ethically sound. By adhering to these standards, countries can ensure that AI technologies are consistent and interoperable, promoting innovation while safeguarding public interests.

Data privacy and security are major concerns that require global cooperation. With AI systems relying on vast amounts of data, ensuring this data is protected and used ethically is a significant challenge. International agreements and regulations, such as the General Data Protection Regulation (GDPR) in the European Union, set high standards for data protection. However, for these regulations to be truly effective, there must be cooperation between countries to ensure that data protection measures are consistent and enforceable across borders. This cooperation helps prevent data breaches and misuse, protecting individuals' privacy rights globally.

Ethical considerations are at the heart of global cooperation in AI. Different cultures and societies have varying perspectives on ethics, and it is essential to find common ground in these diverse viewpoints. International forums and conferences, such as the World Summit on the Information Society (WSIS) and the United Nations' AI for Good Global Summit, provide platforms for stakeholders from different regions to discuss and align on ethical principles for AI. These discussions help to create a global consensus on ethical AI practices, ensuring that AI technologies respect human rights, promote fairness, and prevent harm.

Another critical aspect of global cooperation is addressing the socio-economic impacts of AI. The deployment of AI has the potential to significantly disrupt labor markets, creating challenges such as job displacement and economic inequality. Countries must work together to develop strategies that mitigate these impacts and ensure that the benefits of AI are equitably distributed. This includes sharing best practices for workforce retraining, implementing social safety nets, and fostering inclusive growth. By collaborating on these issues, countries can create a more resilient global economy that adapts to the transformative effects of AI.

Research and development in AI also benefit greatly from global cooperation. Collaborative research initiatives, such as the Partnership on AI and the AI4EU project, bring together researchers, companies, and governments to share knowledge, resources, and expertise. These collaborations accelerate the advancement of AI technologies and ensure that innovations are guided by ethical considerations. Additionally, international cooperation in research helps to address global challenges, such as climate change, public health, and sustainable development, by leveraging AI to find innovative solutions.

Cybersecurity is another area where global cooperation is essential. AI systems are vulnerable to cyber-attacks, and ensuring their security requires coordinated efforts across countries. International cooperation in cybersecurity involves sharing threat intelligence, developing joint defense strategies, and establishing protocols for responding to cyber incidents. This collective approach helps to protect critical infrastructure, safeguard sensitive data, and maintain the integrity of AI systems.

Regulatory frameworks for AI also benefit from global cooperation. Countries can learn from each other's experiences and adopt best practices in AI regulation. Harmonizing regulations across borders helps to create a level playing field for AI development and deployment, preventing regulatory arbitrage and ensuring that AI technologies meet high standards of safety and ethics. International regulatory bodies, such as the European Union Agency for Cybersecurity (ENISA) and the International Telecommunication Union (ITU), play a crucial role in facilitating this cooperation.

In conclusion, global cooperation is essential for the responsible development and deployment of AI. By working together, countries can establish international standards, protect data privacy and security, address ethical considerations, mitigate socio-economic impacts, foster collaborative research, enhance cybersecurity, and harmonize regulatory frameworks. This cooperation ensures that AI technologies are used ethically and for the benefit of all humanity, addressing the challenges and opportunities presented by this transformative technology in a unified and effective manner.

Chapter 3: The Invaluable Nature of Human Life

Philosophical Foundations

Human Dignity and Ethics

The ethical considerations surrounding artificial intelligence and robotics are deeply rooted in the fundamental concept of human dignity. This principle emphasizes the intrinsic worth of every individual, recognizing that all people deserve respect and protection regardless of their circumstances. As AI technologies become more advanced and pervasive, it is crucial to integrate this core value into their development and deployment, ensuring that human dignity remains at the forefront of ethical AI practices.

AI systems have the potential to profoundly impact human lives, offering benefits such as improved healthcare, enhanced productivity, and greater access to information. However, these technologies also pose significant ethical challenges that must be addressed to prevent harm and uphold human dignity. One of the primary concerns is the potential for AI to infringe on privacy rights. AI systems often rely on vast amounts of data, including personal and sensitive information, to function effectively. Ensuring that this data is collected, stored, and used in ways that respect individual privacy is essential. This involves implementing robust data protection measures and providing individuals with control over their personal information.

Another critical aspect of upholding human dignity in AI is preventing discrimination and bias. AI systems can inadvertently perpetuate existing societal biases if they are trained on biased data or if their algorithms are not carefully designed to detect and mitigate bias. This can result in unfair treatment of certain groups, undermining the principle of equality. Ensuring fairness in AI requires rigorous testing, continuous monitoring, and the implementation of bias-correction techniques. For instance, in hiring processes, AI systems should be evaluated to ensure they do not favor candidates based on gender, race, or other irrelevant factors.

Transparency is also a key factor in maintaining human dignity in AI. Users and stakeholders need to understand how AI systems make decisions, especially when these decisions significantly affect people's lives. This transparency can be achieved through explainable AI, which involves designing systems that can provide clear and understandable explanations for their decisions. When individuals understand the reasoning behind an AI system's actions, they are more likely to trust and accept its outcomes. This trust is crucial for the responsible adoption of AI technologies in various sectors, from healthcare to criminal justice.

Human oversight is another essential element in preserving human dignity in the age of AI. While AI systems can process information and make decisions rapidly, they lack the nuanced understanding and ethical judgment that humans possess. Ensuring that humans remain in control of critical decisions, especially those involving ethical dilemmas, is vital. For example, in medical contexts, AI can assist doctors by providing diagnostic suggestions or treatment options, but the final decision should always rest with a qualified healthcare professional who can consider the broader context and the patient's unique needs.

The principle of human dignity also extends to ensuring that AI technologies are accessible and beneficial to all. This involves addressing the digital divide and ensuring that marginalized and underserved communities can access and benefit from AI advancements. Promoting inclusivity in AI development can help bridge gaps in education, healthcare, and economic opportunities, contributing to a more equitable society. Governments and organizations should work together to provide resources, training, and infrastructure that enable widespread access to AI technologies.

Moreover, the ethical design of AI systems must consider the potential long-term impacts on human dignity. This includes evaluating how AI technologies might shape social interactions, employment, and individual autonomy. For instance, while AI can automate many tasks and improve efficiency, it also raises concerns about job displacement and the devaluation of human labor. Addressing these concerns requires proactive measures such as developing policies for workforce retraining, supporting new job creation, and fostering a societal dialogue about the future role of AI in the workplace.

In conclusion, the ethical deployment of AI and robotics hinges on a steadfast commitment to human dignity. This involves ensuring privacy, preventing discrimination, promoting transparency, maintaining human oversight, and fostering inclusivity. By embedding these values into the core of AI development and regulation, we can harness the transformative potential of AI while safeguarding the intrinsic worth and rights of every individual. As AI continues to evolve, prioritizing human dignity will be essential in guiding its integration into society in ways that are ethical, fair, and beneficial for all.

Historical Perspectives

The development of artificial intelligence and robotics has been profoundly influenced by historical perspectives that have shaped the ethical frameworks guiding these technologies. To fully appreciate the current ethical considerations, it is essential to explore the historical context in which these ideas emerged and evolved. This journey begins with the foundational contributions of Isaac Asimov and extends to contemporary advancements that address the limitations of earlier frameworks.

In the mid-20th century, Isaac Asimov introduced his Three Laws of Robotics, which laid the groundwork for ethical discussions in the field. These laws were groundbreaking at the time, prioritizing human safety and ethical behavior in the development and operation of robots. The First Law, which states that a robot may not harm a human being or, through inaction, allow a human being to come to harm, underscored the paramount importance of human safety. This principle was revolutionary, setting a standard that human life and well-being must be the foremost consideration in robotic ethics.

The Second Law, which mandates that a robot must obey human orders unless such orders conflict with the First Law, introduced the concept of hierarchical obedience. This law ensured that robots would follow human commands, thereby reinforcing human control over robotic systems. However, it also highlighted the potential for ethical dilemmas when human orders were conflicting or irrational. The Third Law, which states that a robot must protect its own existence as long as such protection does not conflict with the First or Second Laws, recognized the importance of self-preservation for functional reliability while ensuring that this did not override the primary ethical imperatives of human safety and obedience.

Asimov's laws were not only a product of their time but also a response to the technological optimism and anxieties that characterized the mid-20th century. The rapid advancements in automation and computing technologies sparked both excitement and fear about the potential and risks of intelligent machines. Asimov's narratives explored these themes, using fictional scenarios to probe the strengths and limitations of his ethical framework. These stories helped to shape public and scientific discourse around robotics, influencing how both laypeople and experts thought about the ethical dimensions of intelligent machines.

However, as technology progressed, the limitations of Asimov's laws became more apparent. The binary nature of the laws did not account for the complexities and nuances of real-world situations. For example, the First Law did not provide guidance on how to balance different types of harm or how to prioritize among multiple humans in danger. Similarly, the Second Law's mandate to obey human orders did not consider the potential for conflicting commands from different humans or the challenge of interpreting ambiguous or irrational orders. The Third Law's focus on self-preservation assumed a relatively simple understanding of robot functionality, which did not align with the sophisticated and interconnected nature of modern AI systems.

The evolving nature of AI and robotics necessitated a more comprehensive and nuanced ethical framework. Contemporary advancements in AI have led to the development of systems that are far more complex and capable than those envisioned in Asimov's time. These advancements have highlighted the need for ethical principles that can address the intricate challenges posed by modern technologies. For instance, the rise of machine learning and neural networks has introduced new dimensions of unpredictability and opacity in AI decision-making processes, necessitating greater transparency and accountability.

Moreover, the integration of AI into critical areas such as healthcare, finance, and autonomous transportation has underscored the importance of fairness and equity. Modern ethical frameworks emphasize the need to prevent biases in AI systems, ensuring that these technologies do not perpetuate or exacerbate existing inequalities. This involves rigorous testing, continuous monitoring, and the implementation of bias-correction techniques. Additionally, the ethical use of AI in surveillance and law enforcement requires robust safeguards to protect individual rights and freedoms.

The recognition of these complexities has led to the formulation of more sophisticated ethical guidelines, such as the Seven Directives, which build upon and extend Asimov's original principles. These directives aim to provide a robust framework for the ethical development and deployment of AI and robotics, addressing the limitations of Asimov's laws while responding to contemporary challenges. By emphasizing principles such as human dignity, fairness, transparency, and accountability, these new frameworks seek to ensure that AI technologies serve humanity in the most ethical and beneficial ways possible.

In conclusion, the historical perspectives on AI ethics, from Asimov's pioneering contributions to contemporary advancements, highlight the evolving nature of ethical considerations in this field. As technology continues to advance, it is essential to develop and refine ethical frameworks that can address the complex and nuanced challenges posed by modern

AI and robotics. By learning from the past and adapting to the present, we can create a future where AI technologies are developed and deployed in ways that uphold human dignity, promote fairness, and enhance the well-being of all individuals.

Modern Ethical Theories

The evolution of ethical theories in the realm of artificial intelligence and robotics reflects a journey from simplistic rule-based approaches to more nuanced, comprehensive frameworks designed to address the complexities of modern technology. Understanding these modern ethical theories requires an exploration of their foundations, principles, and applications in today's rapidly advancing technological landscape.

The ethical frameworks for AI and robotics have significantly evolved from the foundational ideas introduced by Isaac Asimov. His Three Laws of Robotics were pioneering at the time, offering a simple yet powerful set of rules to guide the behavior of robots. However, as technology has advanced, these laws have been found insufficient to address the myriad ethical challenges posed by contemporary AI systems. This realization has led to the development of more sophisticated ethical theories that consider the broader implications of AI on society.

One of the prominent modern ethical frameworks is the concept of "beneficence," which emphasizes that AI systems should actively promote the well-being of individuals and society. This principle goes beyond merely preventing harm, as Asimov's First Law suggests, and focuses on the proactive benefits that AI can provide. For example, in healthcare, AI systems designed under the principle of beneficence aim to enhance patient care, improve diagnostic accuracy, and facilitate personalized treatment plans. This proactive approach ensures that AI technologies contribute positively to human health and well-being.

Another crucial modern ethical theory is "justice," which requires that AI systems operate fairly and equitably. This principle addresses the potential for bias and discrimination in AI algorithms, ensuring that all individuals are treated fairly regardless of their background. Implementing justice in AI involves rigorous testing and validation processes to detect and mitigate biases in data and algorithms. For instance, in hiring algorithms, the principle of justice mandates that these systems do not discriminate against candidates based on race, gender, or other protected characteristics, promoting equal opportunities for all applicants.

The principle of "autonomy" is also central to modern ethical theories in AI. Autonomy respects the right of individuals to make informed decisions about their own lives. In the context of AI, this means ensuring that users have control over how AI systems interact with them and their data. Transparent algorithms and clear communication about how AI systems work are essential to uphold this principle. For example, in consumer applications, AI systems should provide users with clear explanations of their decisions and allow them to opt-out or modify how their data is used, ensuring that user autonomy is respected.

"Non-maleficence," which means "do no harm," is another fundamental ethical principle that modern AI frameworks emphasize. While similar to Asimov's First Law, non-maleficence in modern contexts encompasses a broader range of considerations, including psychological and societal harm. This principle requires AI developers to anticipate and mitigate potential negative impacts of their technologies. For example, social media algorithms designed to maximize user engagement must also consider the potential for addiction, mental health issues, and the spread of misinformation, ensuring that their operation does not inadvertently cause harm.

The principle of "transparency" is critical in modern ethical theories, addressing the need for openness in AI development and deployment. Transparency involves making the decision-making processes of AI systems understandable to users and stakeholders. This is particularly important in sectors where AI decisions can have significant impacts, such as criminal justice or finance. Transparent algorithms help build trust and allow for accountability, as stakeholders can understand, evaluate, and challenge AI decisions if necessary.

"Accountability" ensures that there are mechanisms in place to hold AI developers and users responsible for the outcomes of their systems. This principle requires clear frameworks for liability and redress, ensuring that those affected by AI decisions have avenues to seek justice. For example, if an autonomous vehicle is involved in an accident, accountability frameworks determine who is responsible and how victims can be compensated, ensuring that ethical standards are upheld.

The concept of "sustainability" has also emerged as a key consideration in modern ethical theories. Sustainability focuses on the long-term impacts of AI technologies on the environment and future generations. This principle encourages the development of AI systems that are energy-efficient and environmentally friendly, promoting sustainable practices in the technology sector. For instance, AI models designed for climate prediction and environmental monitoring must operate in ways that minimize their ecological footprint while maximizing their contributions to sustainability goals.

In conclusion, modern ethical theories in AI and robotics build upon the foundational ideas of Asimov's Three Laws while addressing their limitations through more comprehensive and nuanced principles. These theories emphasize beneficence, justice, autonomy, non-maleficence, transparency, accountability, and sustainability, ensuring that AI technologies are developed and deployed in ways that promote the well-being of individuals and society. By integrating these principles into the core of AI development, we can create technologies that are not only innovative but also ethical, fair, and beneficial for all.

Practical Applications

AI in Healthcare Decisions

The integration of artificial intelligence in healthcare decisions has revolutionized the medical field, offering unprecedented opportunities for improving patient outcomes, enhancing diagnostic accuracy, and streamlining healthcare delivery. AI's ability to process vast amounts of data and identify patterns that may not be evident to human clinicians makes it an invaluable tool in modern medicine. However, the incorporation of AI into healthcare also raises significant ethical and practical considerations that must be carefully navigated to ensure that these technologies serve humanity's best interests.

One of the most transformative applications of AI in healthcare is in diagnostics. AI algorithms can analyze medical images, such as X-rays, MRIs, and CT scans, with remarkable accuracy, often detecting anomalies that might be missed by the human eye. For example, AI systems have been developed to identify early signs of diseases like cancer, potentially enabling earlier intervention and better patient outcomes. In practice, this involves feeding large datasets of medical images into machine learning models, which then learn to recognize the subtle differences between healthy and diseased tissues. These AI-driven diagnostic tools can assist radiologists by providing a second opinion, thereby reducing diagnostic errors and improving the accuracy of medical diagnoses.

In addition to diagnostics, AI is being used to personalize treatment plans. Personalized medicine involves tailoring medical treatment to the individual characteristics of each patient, taking into account factors such as genetics, lifestyle, and environment. AI systems can analyze data from various sources, including electronic health records (EHRs), genetic information, and wearable devices, to predict how different patients will respond to specific treatments. For instance, in oncology, AI can help oncologists choose the most effective chemotherapy regimen for a particular patient based on their genetic profile and the characteristics of their tumor. This personalized approach not only increases the efficacy of treatments but also minimizes adverse effects, enhancing the overall quality of care.

AI is also playing a crucial role in optimizing hospital operations and improving the efficiency of healthcare delivery. Predictive analytics can forecast patient admissions, allowing hospitals to better manage their resources and staff. For example, AI algorithms can predict the likelihood of a patient being readmitted to the hospital after discharge, enabling healthcare providers to implement preventative measures and follow-up care plans. Additionally, AI can streamline administrative tasks, such as scheduling appointments and managing billing, freeing up healthcare professionals to focus more on patient care.

Despite these advancements, the use of AI in healthcare decisions comes with several ethical challenges that must be addressed to ensure its responsible deployment. One of the primary concerns is the issue of data privacy and security. AI systems require access to vast amounts of personal health data to function effectively, raising concerns about the potential for data breaches and misuse. Ensuring that patient data is protected and that privacy is maintained is paramount. This involves implementing robust data encryption methods, securing data storage, and ensuring compliance with regulations such as the Health Insurance Portability and Accountability Act (HIPAA).

Another significant ethical consideration is the potential for bias in AI algorithms. If the data used to train AI systems is not representative of the diverse patient population, the resulting algorithms may produce biased outcomes. For example, an AI system trained primarily on data from one demographic group may not perform as well for patients from different ethnic or socio-economic backgrounds. Addressing this issue requires careful curation of training datasets to ensure diversity and inclusivity, as well as ongoing monitoring and evaluation to detect and mitigate biases in AI systems.

Transparency and explainability are also crucial in the ethical deployment of AI in healthcare. Clinicians and patients need to understand how AI systems arrive at their decisions, especially in critical areas such as diagnosis and treatment planning. This transparency builds trust and allows for informed decision-making. Explainable AI techniques aim to make the decision-making processes of AI systems more transparent, providing clear and understandable explanations for their recommendations. For instance, if an AI system suggests a particular treatment plan, it should be able to explain the factors that influenced its recommendation, such as patient history and clinical guidelines.

Human oversight is essential to ensure that AI systems complement, rather than replace, the expertise of healthcare professionals. While AI can provide valuable insights and support, the final decision should always rest with a qualified clinician who can consider the broader context and individual patient needs. This collaborative approach ensures that AI enhances, rather than diminishes, the quality of care provided to patients.

In conclusion, the integration of AI into healthcare decisions holds great promise for improving patient outcomes, personalizing treatment, and optimizing healthcare delivery. However, realizing these benefits requires careful consideration of ethical and practical challenges. Ensuring data privacy, addressing biases, promoting transparency, and maintaining human oversight are critical to the responsible deployment of AI in healthcare. By navigating these challenges thoughtfully, we can harness the power of AI to enhance healthcare while safeguarding the dignity and well-being of patients.

Ethical Military Robotics

The integration of artificial intelligence in military robotics has revolutionized warfare and defense strategies, offering capabilities that were once the realm of science fiction. However, the deployment of AI in military contexts raises profound ethical questions that must be addressed to ensure that these technologies are used responsibly and justly. Ethical military robotics must navigate a complex landscape where the imperatives of national security intersect with the principles of human rights and international law.

One of the primary ethical concerns in the development and deployment of military AI is the principle of distinction, which is central to international humanitarian law. This principle requires that combatants distinguish between military targets and civilian objects, ensuring that operations are directed only against lawful military targets. AI systems, particularly autonomous weapons, must be designed to accurately identify and engage targets while minimizing harm to civilians. This involves sophisticated algorithms capable of processing vast amounts of sensory data to make split-second decisions in dynamic and chaotic environments. Ensuring the accuracy and reliability of these systems is crucial to prevent unintended civilian casualties and maintain compliance with international law.

Another critical ethical consideration is proportionality, which mandates that the anticipated military advantage of an attack must outweigh the potential harm to civilians and civilian infrastructure. AI systems in military applications must be programmed to assess the proportionality of their actions, weighing the strategic benefits against the humanitarian costs. This requires complex decision-making capabilities and robust ethical guidelines embedded within the AI algorithms. For instance, an autonomous drone conducting a strike must evaluate the potential collateral damage and decide whether the mission's objectives justify the risks to nearby civilians.

The principle of accountability is also paramount in the ethical deployment of military AI. One of the significant challenges posed by autonomous systems is determining responsibility when things go wrong. If an AI-controlled weapon system makes a mistake, it can be difficult to assign blame or seek redress. Ethical frameworks for military AI must include clear accountability mechanisms that define who is responsible for the actions of autonomous systems. This includes ensuring that human operators retain meaningful oversight and control over AI systems, even in highly automated environments. Establishing legal and procedural norms for accountability helps maintain trust in the use of AI in military operations and ensures that there are avenues for addressing grievances and mistakes.

Transparency is essential for building trust and ensuring the ethical use of AI in military contexts. This involves making the decision-making processes of AI systems understandable to human operators and, where appropriate, to the public and international community. Transparency helps to build trust in the reliability and ethical standards of military AI and facilitates oversight by both military authorities and independent bodies. For example, if an autonomous system is involved in an incident, transparent reporting and investigation processes can help determine the facts and ensure accountability.

Ethical military robotics also must address the risk of escalation and unintended consequences. Autonomous weapons have the potential to make rapid decisions that could escalate conflicts more quickly than human operators might intend. Ensuring that AI systems are programmed with robust safeguards to prevent unintended escalation is crucial. This includes implementing protocols that allow for human intervention and de-escalation in real-time. The strategic use of AI in military contexts must consider not only immediate tactical advantages but also long-term strategic stability and the prevention of unnecessary conflict.

The ethical deployment of AI in military robotics also requires a commitment to international cooperation and compliance with global norms. Countries developing and deploying military AI must engage in dialogue with international partners to establish norms and agreements that govern the use of these technologies. This includes participating in international treaties and agreements that regulate the development and use of autonomous weapons and other AI-enabled military systems. International cooperation helps to ensure that AI technologies are used in ways that are consistent with global standards of human rights and humanitarian law.

Furthermore, the integration of ethical considerations into the design and development of military AI systems is essential. This involves interdisciplinary collaboration between technologists, ethicists, legal experts, and military professionals. By incorporating diverse perspectives and expertise, developers can create AI systems that are not only effective but also aligned with ethical and legal standards. For instance, involving ethicists in the design process can help identify potential ethical dilemmas and develop solutions that respect human rights and international law.

In conclusion, the ethical deployment of AI in military robotics requires a comprehensive approach that addresses the principles of distinction, proportionality, accountability, transparency, and the prevention of escalation. Ensuring that AI systems are designed and used in ways that comply with international humanitarian law and respect human rights is essential for maintaining the legitimacy and moral integrity of military operations. Through robust ethical frameworks, international cooperation, and interdisciplinary collaboration, we can harness the capabilities of AI in military contexts while upholding the highest standards of ethics and justice.

AI in Emergency Management

Artificial intelligence has become an integral part of emergency management, providing advanced tools and capabilities that significantly enhance the efficiency and effectiveness of disaster response and recovery efforts. The deployment of AI in this field involves using machine learning algorithms, predictive analytics, and real-time data processing to assist in various stages of emergency management, from preparedness to response and recovery. However, the integration of AI in emergency management also raises important ethical considerations that must be addressed to ensure that these technologies are used responsibly and equitably.

One of the primary applications of AI in emergency management is in the prediction and early warning of natural disasters. Machine learning algorithms can analyze vast amounts of data from various sources, such as satellite imagery, weather sensors, and historical records, to identify patterns and predict potential disasters like hurricanes, earthquakes, and floods. For instance, AI systems can analyze atmospheric data to predict the path and intensity of hurricanes, allowing authorities to issue timely warnings and prepare evacuation plans. The ability to predict disasters with greater accuracy helps mitigate the impact on communities, saving lives and reducing property damage.

In addition to prediction, AI plays a crucial role in the real-time monitoring and assessment of ongoing disasters. Drones equipped with AI-powered image recognition technology can provide real-time aerial footage of disaster-affected areas, helping emergency responders assess the extent of damage and identify the most affected regions. This real-time data allows for more informed decision-making and efficient allocation of resources. For example, during wildfires, AI can analyze data from drones and satellites to map the fire's spread, predict its path, and suggest the best strategies for containment and evacuation.

AI is also instrumental in optimizing resource allocation and logistics during disaster response. In the immediate aftermath of a disaster, ensuring that resources such as food, water, medical supplies, and personnel are distributed efficiently is critical. AI algorithms can analyze data on population density, infrastructure damage, and road conditions to develop optimal routes for delivering supplies and deploying emergency teams. This ensures that aid reaches those in need as quickly as possible, minimizing delays and improving the overall effectiveness of the response effort.

Furthermore, AI can enhance communication and coordination among emergency responders. Natural language processing (NLP) algorithms can analyze social media posts, news reports, and emergency calls to identify urgent needs and emerging issues in real-time. This information can be used to coordinate efforts among different agencies and ensure that resources are directed where they are most needed. For instance, during a major flood, NLP algorithms can monitor social media for reports of stranded individuals or areas experiencing severe flooding, allowing emergency responders to prioritize their efforts accordingly.

Despite the numerous benefits of AI in emergency management, there are significant ethical considerations that must be addressed. One of the main concerns is the potential for bias in AI algorithms. If the data used to train AI systems is not representative of all communities, the resulting predictions and decisions may disproportionately benefit some groups while neglecting others. For example, if AI algorithms are trained primarily on data from urban areas, they may be less accurate in predicting and responding to disasters in rural or underserved regions. Addressing this issue requires ensuring that training data is diverse and inclusive, representing a wide range of communities and scenarios.

Privacy is another critical ethical consideration. The use of AI in emergency management often involves the collection and analysis of large amounts of personal data, such as location information, social media activity, and health records. Ensuring that this data is collected and used in a way that respects individuals' privacy rights is essential. This involves implementing robust data protection measures, such as encryption and anonymization, and ensuring that data is used only for legitimate emergency management purposes.

Transparency and accountability are also vital. The decision-making processes of AI systems used in emergency management should be transparent and understandable to ensure that stakeholders can trust and verify the actions taken. This includes providing clear explanations of how AI algorithms arrive at their predictions and decisions and ensuring that there are mechanisms in place for accountability if errors or biases occur. For instance, if an AI system incorrectly predicts the path of a hurricane, leading to inadequate preparation and increased damage, it is important to investigate and address the factors that contributed to the error.

Finally, the deployment of AI in emergency management should be guided by the principle of equity. It is crucial to ensure that the benefits of AI are distributed fairly and that all communities, especially those that are most vulnerable, have access to these technologies. This involves prioritizing the needs of marginalized and underserved populations and ensuring that AI systems are designed and implemented in ways that address their specific challenges and needs.

In conclusion, AI has the potential to significantly enhance emergency management by improving disaster prediction, real-time monitoring, resource allocation, and communication. However, the ethical considerations surrounding its use must be carefully addressed to ensure that these technologies are used responsibly, equitably, and in ways that respect privacy and human rights. By navigating these challenges thoughtfully, we can harness the power of AI to build more resilient communities and save lives in the face of disasters.

Decision-Making Algorithms

Value-based Algorithms

The development and implementation of value-based algorithms represent a significant advancement in the field of artificial intelligence, offering a framework that integrates ethical values directly into the decision-making processes of AI systems. These algorithms are designed to prioritize human-centric values, ensuring that AI technologies not only perform efficiently but also align with societal norms and ethical standards. The adoption of value-based algorithms is crucial for fostering trust and ensuring that AI serves the best interests of humanity.

Value-based algorithms are structured to incorporate ethical principles such as fairness, accountability, and transparency into their core functionalities. Unlike traditional algorithms that primarily focus on optimizing specific performance metrics, value-based algorithms are programmed to balance these metrics with ethical considerations. This holistic approach ensures that AI systems operate in ways that are socially beneficial and ethically sound.

One of the primary areas where value-based algorithms are making a significant impact is in healthcare. AI systems equipped with these algorithms can assist in diagnosing diseases, recommending treatments, and managing patient care, all while ensuring that ethical principles such as patient autonomy, privacy, and fairness are upheld. For instance, an AI system designed to recommend treatment plans must consider not only the clinical effectiveness of different options but also the patient's personal preferences, cultural values, and potential socioeconomic impacts. By integrating these factors, value-based algorithms help ensure that healthcare decisions are patient-centered and ethically responsible.

In the context of autonomous vehicles, value-based algorithms play a critical role in enhancing safety and ethical decision-making. Autonomous driving systems must navigate complex and dynamic environments where they encounter various ethical dilemmas. For example, in scenarios where a collision is unavoidable, these algorithms must make split-second decisions that balance the safety of passengers, pedestrians, and other road users. Value-based algorithms are programmed to prioritize human life and minimize harm, considering the ethical implications of their actions. This approach ensures that autonomous vehicles operate in ways that are not only safe but also ethically defensible.

Another important application of value-based algorithms is in the field of finance. AI systems used in financial services must make decisions that affect people's livelihoods and economic well-being. For example, algorithms used in loan approval processes must ensure that decisions are fair and unbiased, taking into account the potential for discrimination based on race, gender, or socioeconomic status. By incorporating fairness into their decision-making criteria, value-based algorithms help prevent discriminatory practices and promote financial inclusion.

The development of value-based algorithms involves several key steps. First, it is essential to identify the specific values and ethical principles that should guide the algorithm's decisions. This process typically involves input from a diverse group of stakeholders, including ethicists, domain experts, and representatives from affected communities. Once the relevant values have been identified, they are translated into quantifiable metrics that can be incorporated into the algorithm's design. This translation process is complex and requires careful consideration to ensure that the chosen metrics accurately reflect the underlying ethical principles.

Next, the algorithm is trained using a combination of traditional performance data and value-based criteria. This dual training approach ensures that the algorithm not only performs well according to conventional metrics but also adheres to ethical standards. During this phase, it is crucial to continuously monitor and evaluate the algorithm's performance, making adjustments as needed to address any emerging ethical concerns or biases.

The implementation of value-based algorithms also requires robust governance frameworks to ensure ongoing oversight and accountability. This includes establishing clear protocols for monitoring the algorithm's decisions, assessing their ethical implications, and making necessary adjustments. Additionally, transparency is vital to building trust and ensuring that stakeholders understand how the algorithm operates and makes decisions. Providing clear, understandable explanations of the algorithm's decision-making processes helps build confidence and allows for meaningful stakeholder engagement.

Despite the significant benefits of value-based algorithms, there are also challenges that must be addressed. One of the primary challenges is the potential for conflicts between different values. For example, in healthcare, there may be situations where the value of patient autonomy conflicts with the value of maximizing clinical outcomes. Addressing these conflicts requires careful balancing and prioritization of values, guided by ethical principles and stakeholder input.

Another challenge is the potential for bias in the algorithm's design and implementation. Ensuring that value-based algorithms are fair and unbiased requires rigorous testing and continuous monitoring to identify and mitigate any biases that may arise. This involves using diverse and representative datasets during the training process and implementing robust mechanisms for detecting and addressing biases.

In conclusion, value-based algorithms represent a significant advancement in AI, integrating ethical considerations directly into the decision-making processes of AI systems. By prioritizing values such as fairness, accountability, and transparency, these algorithms ensure that AI technologies operate in ways that are socially beneficial and ethically sound. The development and implementation of value-based algorithms require careful consideration of ethical principles, stakeholder input, and robust governance frameworks to ensure ongoing oversight and accountability. Despite the challenges, value-based algorithms hold great promise for enhancing the ethical and social impact of AI, fostering trust, and ensuring that AI serves the best interests of humanity.

Risk Assessment Models

The integration of value-based algorithms into risk assessment models has transformed various sectors by enhancing decision-making processes with ethical and human-centric considerations. These algorithms, designed to prioritize human values and ethical principles, are increasingly being applied in scenarios where assessing and managing risk is crucial. From finance and healthcare to environmental management and beyond, value-based risk assessment models ensure that decisions are not only efficient but also ethically sound and aligned with societal values.

In the financial sector, risk assessment is a fundamental aspect of operations, from evaluating loan applications to managing investment portfolios. Traditional risk assessment models primarily focus on financial metrics such as credit scores, income levels, and market trends. However, value-based algorithms incorporate ethical considerations into these assessments. For instance, when evaluating loan applications, these algorithms consider not only the applicant's creditworthiness but also factors such as potential biases and the broader social impact of lending decisions. By doing so, they promote financial inclusion and fairness, ensuring that underrepresented and underserved communities are not unfairly disadvantaged.

In healthcare, risk assessment models are critical for patient diagnosis, treatment planning, and management. Value-based algorithms enhance these models by integrating ethical principles such as patient autonomy, fairness, and the right to privacy. For example, when determining the risk of a patient developing a particular condition, these algorithms consider diverse data sources, including genetic information, lifestyle factors, and environmental exposures. They also ensure that data privacy is maintained and that patients' personal information is used responsibly. This comprehensive approach not only improves the accuracy of risk predictions but also ensures that healthcare decisions respect patients' rights and values.

Environmental management is another area where value-based risk assessment models play a pivotal role. These models are used to evaluate the potential impact of environmental policies, industrial activities, and natural disasters on ecosystems and human communities. Value-based algorithms enhance these assessments by incorporating ethical considerations such as sustainability, equity, and the precautionary principle. For instance, when assessing the risk of an industrial project, these algorithms consider its potential environmental impact, the fairness of its benefits and burdens across different communities, and the long-term sustainability of natural resources. This ensures that environmental decisions are made in a way that protects the planet and promotes social justice.

The development and implementation of value-based risk assessment models involve several key steps. First, it is essential to identify the specific values and ethical principles that should guide the assessment process. This requires input from a diverse group of stakeholders, including ethicists, domain experts, and representatives from affected communities. Once these values have been identified, they are translated into quantifiable metrics that can be incorporated into the risk assessment model. This translation process is complex and requires careful consideration to ensure that the chosen metrics accurately reflect the underlying ethical principles.

Next, the risk assessment model is developed using a combination of traditional data sources and value-based criteria. This dual approach ensures that the model not only performs well according to conventional risk metrics but also adheres to ethical standards. During this phase, it is crucial to continuously monitor and evaluate the model's performance, making adjustments as needed to address any emerging ethical concerns or biases. For example, if the model consistently underestimates risks for certain vulnerable populations, adjustments must be made to ensure that these groups are adequately protected.

The implementation of value-based risk assessment models also requires robust governance frameworks to ensure ongoing oversight and accountability. This includes establishing clear protocols for monitoring the model's decisions, assessing their ethical implications, and making necessary adjustments. Additionally, transparency is vital to building trust and ensuring that stakeholders understand how the model operates and makes decisions. Providing clear, understandable explanations of the model's decision-making processes helps build confidence and allows for meaningful stakeholder engagement.

Despite the significant benefits of value-based risk assessment models, there are also challenges that must be addressed. One of the primary challenges is the potential for conflicts between different values. For instance, in healthcare, there may be situations where the value of patient autonomy conflicts with the need to maximize clinical outcomes. Addressing these conflicts requires careful balancing and prioritization of values, guided by ethical principles and stakeholder input.

Another challenge is the potential for bias in the model's design and implementation. Ensuring that value-based risk assessment models are fair and unbiased requires rigorous testing and continuous monitoring to identify and mitigate any biases that may arise. This involves using diverse and representative datasets during the training process and implementing robust mechanisms for detecting and addressing biases.

In conclusion, value-based algorithms significantly enhance risk assessment models by integrating ethical considerations into decision-making processes. By prioritizing values such as fairness, accountability, and sustainability, these models ensure that risk assessments are not only accurate but also ethically sound and aligned with societal values. The development and implementation of value-based risk assessment models require careful consideration of ethical principles, stakeholder input, and robust governance frameworks to ensure ongoing oversight and accountability. Despite the challenges, these models hold great promise for improving decision-making across various sectors, fostering trust, and ensuring that AI serves the best interests of humanity.

Case Studies in Ethical Decision-Making

The application of artificial intelligence in ethical decision-making is best illustrated through real-world case studies that highlight the challenges and successes of integrating AI into various sectors. These case studies provide valuable insights into how ethical principles are applied in practice, demonstrating the complexities and nuances involved in ensuring that AI technologies align with societal values and ethical standards.

One notable case study is the deployment of AI in the criminal justice system, specifically in risk assessment algorithms used to predict the likelihood of reoffending. These algorithms, such as COMPAS (Correctional Offender Management Profiling for Alternative Sanctions), are designed to assist judges in making more informed decisions about bail, sentencing, and parole. The ethical challenge here lies in ensuring that the algorithms do not perpetuate or exacerbate existing biases in the justice system. Studies have shown that some risk assessment tools have exhibited racial biases, leading to higher risk scores for minority groups compared to their white counterparts. Addressing these biases requires rigorous testing and continuous monitoring of the algorithms, as well as incorporating fairness as a core principle in their design. By doing so, the justice system can work towards more equitable outcomes while leveraging the predictive power of AI.

In healthcare, AI's role in diagnostic decision-making presents another compelling case study. AI systems like IBM Watson have been used to analyze medical records, research papers, and clinical trial data to provide diagnostic recommendations and treatment plans. The ethical considerations in this context revolve around patient autonomy, data privacy, and the accuracy of AI recommendations. For instance, in one case, an AI system misdiagnosed a patient's condition due to a lack of comprehensive data, highlighting the importance of high-quality, diverse datasets in training AI models. Ensuring that patients' privacy is protected and that they retain control over their personal health information is also crucial. Moreover, healthcare providers must be transparent about the AI's role in the decision-making process and ensure that the final decisions are made by qualified medical professionals who can consider the AI's recommendations in the broader context of each patient's unique circumstances.

The use of AI in autonomous vehicles provides another rich case study in ethical decision-making. Autonomous vehicles must navigate complex ethical dilemmas, such as the trolley problem, where the AI must choose between two unfavorable outcomes in an unavoidable accident. For example, should an autonomous vehicle prioritize the safety of its passengers over pedestrians? The ethical framework for these decisions involves programming the AI with principles that prioritize minimizing harm and protecting human life. In practice, companies developing autonomous vehicles, such as Tesla and Waymo, conduct extensive simulations and real-world testing to refine their decision-making algorithms. They also engage with ethicists, regulators, and the public to develop guidelines that balance safety, legal, and ethical considerations.

AI's role in financial decision-making, particularly in credit scoring and loan approval, presents another case study. Algorithms used by financial institutions to assess creditworthiness must be designed to avoid discrimination and ensure fairness. In one instance, an AI system used by a major bank was found to disproportionately deny loans to minority applicants, raising concerns about bias in the training data and algorithm design. To address this, financial institutions are adopting value-based algorithms that incorporate fairness and transparency into their decision-making processes. This involves using diverse datasets, implementing bias detection and mitigation techniques, and providing clear explanations for decisions to ensure that all applicants are treated equitably.

Environmental management and disaster response also benefit from AI-driven ethical decision-making. AI systems can predict natural disasters, optimize resource allocation, and coordinate emergency responses. For instance, during the California wildfires, AI was used to analyze satellite imagery and predict the fire's spread, helping firefighters allocate resources more effectively. The ethical considerations here include ensuring that the AI's predictions are accurate and reliable, protecting the privacy of individuals whose data is used in the analysis, and prioritizing the safety and well-being of affected communities. Transparent communication about the AI's role and the limitations of its predictions is essential to maintain public trust and facilitate informed decision-making.

In conclusion, these case studies illustrate the critical importance of integrating ethical principles into AI decision-making processes across various sectors. By prioritizing fairness, transparency, accountability, and human-centric values, AI technologies can be designed and implemented in ways that enhance societal well-being while respecting ethical standards. These examples highlight the ongoing challenges and opportunities in ensuring that AI serves humanity in the most ethical and beneficial ways possible. Through continuous evaluation, stakeholder engagement, and adherence to robust ethical frameworks, the potential of AI can be harnessed responsibly and equitably.

Chapter 4: AI and Robotic Self-Preservation

Importance of Resilience

Building Robust AI Systems

Building robust AI systems is a multifaceted endeavor that requires careful consideration of technical, ethical, and practical dimensions to ensure that these systems are reliable, safe, and aligned with human values. The development of robust AI involves creating systems that can operate effectively under a wide range of conditions, handle unexpected inputs gracefully, and recover from errors or failures without causing harm. This comprehensive approach is essential for fostering trust and ensuring that AI technologies can be integrated into critical aspects of society.

One of the key components of building robust AI systems is the implementation of resilience. Resilience in AI refers to the system's ability to maintain functionality and recover quickly from disruptions or failures. This involves designing algorithms that can handle noise and variability in the data, as well as incorporating redundancy and fault-tolerant mechanisms. For instance, in autonomous vehicles, resilience is achieved by using multiple sensors and redundant systems that can take over if one component fails. This ensures that the vehicle can continue to operate safely even in the event of a hardware malfunction.

Another crucial aspect of robustness is adaptability. AI systems must be able to adapt to new and changing environments without significant degradation in performance. This requires the ability to learn from new data and experiences, updating models and decision-making processes accordingly. Machine learning techniques such as reinforcement learning and continual learning are particularly valuable in this context, as they enable AI systems to refine their behavior over time based on feedback and new information. For example, AI-driven medical diagnostic tools must continuously learn from new patient data and medical research to improve their accuracy and relevance.

Robust AI systems also require rigorous testing and validation to ensure their reliability and safety. This involves extensive simulation and real-world testing under diverse conditions to identify potential weaknesses and areas for improvement. For example, before deploying an AI system in a healthcare setting, it must be tested on a wide range of medical cases to ensure that it can accurately diagnose and recommend treatments for various conditions. Additionally, validation processes must include edge cases and rare scenarios to ensure that the system can handle unexpected situations appropriately.

Ethical considerations are integral to building robust AI systems. Ensuring that AI operates in ways that align with human values and ethical principles is paramount. This includes addressing issues such as bias, fairness, transparency, and accountability. Bias in AI can lead to unfair and discriminatory outcomes, undermining trust and causing harm. Therefore, robust AI systems must be designed to detect and mitigate biases in the data and algorithms. This involves using diverse and representative datasets, implementing fairness-aware machine learning techniques, and conducting regular audits to identify and address potential biases.

Transparency and explainability are also critical for robust AI. Users and stakeholders must understand how AI systems make decisions, especially in high-stakes domains such as healthcare, finance, and criminal justice. Explainable AI techniques provide insights into the decision-making processes of AI models, helping users understand the rationale behind predictions and recommendations. This transparency fosters trust and allows for meaningful human oversight and intervention when necessary. For instance, in financial services, AI systems used for credit scoring must provide clear explanations for their decisions to ensure that applicants can understand and contest any adverse outcomes.

Accountability mechanisms are essential to ensure that AI systems operate responsibly and that there are clear processes for addressing any issues that arise. This includes establishing governance frameworks that define roles and responsibilities for AI development, deployment, and monitoring. Robust AI systems must include mechanisms for tracking decisions and actions, enabling audits and investigations when necessary. For example, in autonomous vehicles, accountability frameworks must define who is responsible in the event of an accident and ensure that there are protocols for investigating and addressing the cause of the incident.

Security is another critical aspect of robustness in AI systems. Ensuring that AI systems are secure from cyber threats and unauthorized access is vital to maintaining their integrity and reliability. This involves implementing robust cybersecurity measures, such as encryption, access controls, and regular security audits. AI systems must also be designed to detect and respond to security breaches, protecting sensitive data and preventing malicious exploitation. For instance, AI systems used in critical infrastructure must be safeguarded against cyber-attacks that could disrupt essential services.

In conclusion, building robust AI systems requires a holistic approach that integrates technical resilience, adaptability, rigorous testing, and ethical considerations. By prioritizing reliability, safety, and alignment with human values, we can develop AI technologies that are capable of operating effectively in diverse and dynamic environments. Ensuring transparency, accountability, and security further strengthens the robustness of AI systems, fostering trust and enabling their integration into critical aspects of society. As AI continues to advance, the principles and practices of robust AI development will be essential for harnessing its full potential in a responsible and beneficial manner.

Redundancy and Fail-safes

The development of redundancy and fail-safes in AI systems is crucial to ensuring their robustness and reliability. As AI technologies become more integrated into critical aspects of society, from healthcare to transportation to finance, the ability to prevent, detect, and recover from failures is essential for maintaining trust and safety. Redundancy and fail-safes are foundational elements in designing AI systems that can handle unexpected disruptions and continue to function effectively.

Redundancy in AI systems involves incorporating multiple layers of backup components and systems to ensure that if one part fails, others can take over without compromising the overall functionality. This approach is widely used in various industries to enhance system reliability and safety. For instance, in autonomous vehicles, redundancy is achieved by using multiple sensors, such as cameras, lidar, and radar, to perceive the environment. If one sensor fails or provides inaccurate data, the other sensors can compensate, ensuring that the vehicle can still navigate safely.

In healthcare, redundancy is vital for medical AI systems that assist in diagnostics and treatment planning. These systems often rely on multiple sources of data, such as electronic health records, imaging studies, and laboratory results, to make accurate decisions. By cross-referencing information from different sources, the system can verify the accuracy of its analyses and recommendations. This redundancy helps prevent errors that could arise from relying on a single source of data, thus enhancing the safety and reliability of medical AI applications.

Fail-safes are mechanisms designed to automatically trigger safe modes of operation or shut down the system in case of a malfunction or unexpected event. The primary goal of a fail-safe is to prevent harm and protect both the system and its users. In industrial robotics, fail-safes are implemented to stop robot operations if a human enters the robot's work area, preventing potential accidents. Similarly, in AI-driven financial systems, fail-safes can halt trading activities if unusual patterns are detected, protecting against market anomalies or cyber-attacks.

The implementation of fail-safes in AI systems also involves creating protocols for safe degradation, where the system continues to operate at a reduced capacity rather than shutting down completely. This approach ensures that essential functions are maintained while minimizing the risk of further damage or failure. For example, in autonomous drones used for disaster response, a fail-safe mechanism might allow the drone to continue flying at a lower altitude and speed if it detects a loss of communication with the control center, ensuring that it can still perform critical tasks while mitigating risks.

Building redundancy and fail-safes into AI systems requires a comprehensive understanding of potential failure modes and their impacts. This involves conducting thorough risk assessments and scenario analyses to identify possible points of failure and develop appropriate mitigation strategies. Engineers and developers must consider both hardware and software components, as well as the interactions between them, to create robust systems that can withstand a variety of challenges.

Moreover, redundancy and fail-safes must be designed with ethical considerations in mind. Ensuring that these mechanisms do not introduce new risks or biases is essential for maintaining fairness and trust. For instance, in designing fail-safes for autonomous vehicles, developers must ensure that the fail-safe actions do not disproportionately affect certain groups of people or create unintended safety hazards. Ethical considerations also involve transparency in how redundancy and fail-safes are implemented, allowing users and stakeholders to understand the measures in place and trust the system's reliability.

Continuous monitoring and maintenance are also critical to the effectiveness of redundancy and fail-safes. AI systems must be regularly tested and updated to ensure that all components and backup mechanisms are functioning correctly. This involves routine inspections, software updates, and real-world testing to identify and address any emerging issues. In safety-critical applications, such as healthcare and transportation, rigorous regulatory standards and oversight are necessary to ensure that redundancy and fail-safes meet high safety and reliability standards.

In conclusion, redundancy and fail-safes are essential components of building robust AI systems. By incorporating multiple layers of backup components and automatic safety mechanisms, developers can create AI technologies that are reliable, safe, and capable of handling unexpected disruptions. These measures are crucial for maintaining trust and ensuring that AI systems can be integrated into critical aspects of society without compromising safety and ethical standards. As AI continues to evolve, the principles of redundancy and fail-safes will play a vital role in ensuring the resilience and reliability of these transformative technologies.

Case Studies in Resilient AI

The development and deployment of resilient AI systems are illustrated by several case studies that highlight the practical application of redundancy, fail-safes, and robust design principles. These examples demonstrate the importance of ensuring that AI technologies can withstand unexpected disruptions, maintain functionality, and recover gracefully from failures, thereby ensuring reliability and trust in critical applications.

One notable case study is the implementation of AI in autonomous vehicles. Companies like Tesla and Waymo have integrated extensive redundancy and fail-safes into their self-driving car systems to ensure safety and reliability. Autonomous vehicles rely on multiple sensors, including cameras, radar, and lidar, to perceive their environment. This redundancy ensures that if one sensor fails or provides inaccurate data, the others can compensate, allowing the vehicle to continue operating safely. Additionally, these systems include fail-safes that enable the car to perform controlled stops or maneuver to a safe location if a critical failure is detected. This approach minimizes the risk of accidents and enhances passenger safety.

In the healthcare sector, the use of AI in medical diagnostics and treatment planning provides another example of resilient AI systems. IBM Watson, for instance, has been used to assist oncologists in diagnosing and developing treatment plans for cancer patients. The system analyzes vast amounts of medical data, including patient records, clinical trial results, and medical literature, to provide recommendations. To ensure resilience, Watson incorporates multiple layers of data validation and cross-referencing, reducing the likelihood of errors. Additionally, the system is designed to allow human doctors to review and override its recommendations, providing an essential fail-safe that ensures medical decisions are made with human oversight.

Another significant case study is the application of AI in disaster response and management. AI systems are used to predict natural disasters, coordinate emergency responses, and optimize resource allocation. For example, during the California wildfires, AI was employed to analyze satellite imagery and weather data to predict the fire's spread and suggest evacuation routes. These systems are designed with redundancy in data sources, including satellite, drone, and ground-based sensors, to ensure comprehensive and accurate information. Fail-safes are implemented to provide alternative communication and data transmission methods in case primary channels are disrupted. This redundancy and robustness enable more effective and reliable disaster response efforts, ultimately saving lives and resources.

In the financial sector, AI-driven trading platforms and risk management systems illustrate the importance of resilience in AI applications. These systems must process and analyze large volumes of market data in real-time to make informed trading decisions. To ensure reliability, financial AI systems are designed with redundant data feeds and backup algorithms that can take over if primary systems fail. Additionally, fail-safes are implemented to halt trading activities in the event of market anomalies or technical issues, protecting against significant financial losses and maintaining market stability.

The deployment of AI in cybersecurity also highlights the critical role of resilient AI systems. AI is used to detect and respond to cyber threats in real-time, analyzing network traffic, identifying anomalies, and mitigating attacks. These systems are designed with redundancy in detection methods, using a combination of signature-based, anomaly-based, and behavior-based techniques to identify threats. Fail-safes are incorporated to isolate and contain compromised systems, preventing the spread of malware or data breaches. This multi-layered approach ensures that cybersecurity AI systems remain effective and reliable, even in the face of sophisticated and evolving threats.

The integration of AI in industrial automation and robotics further demonstrates the application of resilient AI principles. In manufacturing environments, AI-powered robots and control systems are used to optimize production processes and improve efficiency. These systems are equipped with redundant sensors and control mechanisms to ensure continuous operation. For example, if a robotic arm's primary sensor fails, secondary sensors can take over to maintain precision and functionality. Fail-safes are designed to safely shut down machinery or switch to manual control in the event of critical failures, protecting both human workers and equipment.

In conclusion, these case studies illustrate the essential role of redundancy, fail-safes, and robust design in building resilient AI systems. By integrating multiple layers of backup components, incorporating comprehensive fail-safe mechanisms, and ensuring continuous monitoring and maintenance, these systems can operate reliably under diverse and challenging conditions. These principles not only enhance the safety and effectiveness of AI technologies but also foster trust and confidence in their deployment across critical sectors. As AI continues to advance, the lessons learned from these case studies will be vital in guiding the development of future resilient AI systems that serve humanity ethically and effectively.

Ethical Considerations

Balancing Self-preservation and Human Safety

Balancing self-preservation and human safety in artificial intelligence systems is a complex yet vital task that addresses the fundamental need for AI to protect its functionality without compromising the well-being of humans. This balance is essential in ensuring that AI technologies, which are becoming increasingly integrated into critical areas of our lives, operate reliably and ethically.

In the context of autonomous vehicles, this balance is particularly evident. These vehicles must navigate safely while preserving their operational integrity. Autonomous systems are designed with multiple layers of safety protocols and fail-safes to ensure they can handle unexpected situations. For example, if an autonomous car detects an imminent collision, it must decide how to minimize harm to passengers, pedestrians, and itself. This involves complex algorithms that prioritize human safety over the vehicle's preservation. Engineers achieve this by programming vehicles to follow the principle that the safety of human lives always takes precedence, even if it means damaging the car.

Healthcare robotics also exemplifies the balance between self-preservation and human safety. Medical robots assist in surgeries, patient care, and diagnostics, often operating in high-stakes environments where errors can have severe consequences. These robots must maintain their operational capabilities to provide continuous care, but their design inherently prioritizes patient safety. For instance, surgical robots are equipped with sensors that can detect the slightest resistance or anomalies, prompting the system to pause or adjust its actions to avoid harming the patient. Such systems also have manual override functions that allow human surgeons to take control immediately if any issue arises, ensuring that human oversight can swiftly address potential risks.

The aviation industry's use of AI in flight management systems is another area where balancing self-preservation with human safety is critical. Modern aircraft rely on sophisticated AI systems to manage flight paths, monitor engine performance, and detect potential hazards. These systems are built with redundancies and fail-safes to ensure they can continue operating during technical failures. For instance, if an AI system detects engine failure, it will initiate protocols to stabilize the aircraft and guide it to the nearest safe landing location. The priority is always the safety of passengers and crew, with the system's self-preservation mechanisms serving to support this primary goal.

In the realm of industrial automation, robots are designed to perform tasks that are often hazardous to humans, such as handling toxic materials or working in extreme environments. These robots must be robust enough to withstand harsh conditions while ensuring that their operations do not pose any risk to human workers. Safety protocols include emergency stop functions and isolation zones where robots can perform dangerous tasks without human presence. The systems are programmed to halt operations or switch to a safe mode if they detect any breach of these safety zones, ensuring that human safety is never compromised.

The principles of redundancy and fail-safes are integral to achieving this balance. Redundancy ensures that if one component fails, others can take over, maintaining the system's functionality. This is crucial in life-critical applications, where failure can lead to severe consequences. For example, in autonomous drones used for disaster response, multiple sensors and control systems are employed to ensure that if one fails, others can continue to provide necessary data and control, enabling the drone to complete its mission without endangering lives.

Fail-safes, on the other hand, are designed to bring the system to a safe state in the event of a malfunction. This includes emergency shut-down procedures, safe operational modes, and manual override options. These mechanisms are essential in ensuring that AI systems do not exacerbate dangerous situations. For instance, in nuclear power plants, AI systems monitor reactor conditions and can initiate shutdown procedures if they detect abnormal readings, thus preventing potential disasters.

Ethical considerations play a crucial role in designing these systems. It is not enough for AI to be technically robust; it must also operate within ethical boundaries that prioritize human well-being. This involves continuous assessment and updating of AI systems to ensure they adhere to evolving ethical standards.

Developers and engineers must engage with ethicists, regulators, and the public to understand the broader implications of AI systems and incorporate ethical principles into their design and operation.

In conclusion, balancing self-preservation and human safety in AI systems requires a multi-faceted approach that integrates technical robustness, ethical considerations, and continuous monitoring. Through redundancy and fail-safes, AI systems can achieve the reliability needed to operate in critical environments while prioritizing human safety. This balance is essential in ensuring that AI technologies enhance our lives without posing undue risks, fostering trust and facilitating their integration into society. As AI continues to evolve, maintaining this balance will be fundamental to its responsible and ethical development.

Ethical Frameworks for Self-preservation

Ethical frameworks for self-preservation in AI systems are essential to ensure that while these technologies are designed to protect and maintain their functionality, they do so without compromising human safety or ethical standards. The balance between self-preservation and ethical considerations involves creating AI systems that can operate reliably and safely while prioritizing the well-being of humans.

Self-preservation in AI refers to the system's ability to maintain its operational capabilities, recover from failures, and protect itself from potential threats. This aspect is crucial for ensuring the longevity and reliability of AI systems, especially those deployed in critical environments such as healthcare, transportation, and industrial automation. However, the drive for self-preservation must be carefully managed to prevent any negative impact on human safety or ethical values.

In designing ethical frameworks for AI self-preservation, several key principles must be considered. The first principle is the primacy of human safety. AI systems must be programmed to prioritize human life and well-being over their own operational integrity. For example, in autonomous vehicles, the system should be designed to avoid collisions and protect passengers and pedestrians, even if it means damaging the vehicle itself. This principle ensures that the AI's self-preservation mechanisms do not override the imperative to safeguard human lives.

Another important principle is transparency. AI systems should be designed to operate transparently, providing clear and understandable explanations for their actions, especially in scenarios involving self-preservation. This transparency is crucial for building trust with users and stakeholders, as it allows them to understand how the AI makes decisions and ensures accountability. For instance, in medical AI applications, the system should explain its diagnostic and treatment recommendations, including how it balances patient safety with operational considerations.

Accountability is also a fundamental aspect of ethical self-preservation in AI. There must be clear protocols for holding AI systems and their developers accountable for the decisions and actions of the AI. This includes establishing governance frameworks that define the roles and responsibilities of various stakeholders, including developers, users, and regulatory bodies. In the event of an AI system failure or ethical breach, these protocols ensure that appropriate measures are taken to address the issue and prevent future occurrences.

Redundancy and fail-safes are critical components of an ethical framework for AI self-preservation. Redundancy involves incorporating multiple layers of backup systems and components to ensure that the AI can continue functioning even if one part fails. Fail-safes are mechanisms designed to bring the system to a safe state in the event of a malfunction. These measures are essential for maintaining the reliability and safety of AI systems. For example, in industrial robotics, redundancy might involve using multiple sensors to monitor the robot's environment, while fail-safes could include emergency stop functions that halt operations if a human enters the robot's workspace.

Ethical AI self-preservation also requires continuous monitoring and adaptation. AI systems must be equipped with the capability to continuously monitor their performance and detect any anomalies or potential threats. This involves using advanced analytics and machine learning techniques to identify and address issues before they escalate. Additionally, AI systems should be designed to adapt to changing conditions and learn from new data, ensuring that they can respond effectively to evolving challenges.

Cybersecurity is another critical consideration in the ethical framework for AI self-preservation. AI systems must be protected from cyber threats that could compromise their functionality and safety. This involves implementing robust cybersecurity measures, such as encryption, access controls, and regular security audits, to safeguard the AI system and its data. Ensuring the cybersecurity of AI systems is essential for maintaining their integrity and preventing malicious exploitation.

Finally, the ethical framework for AI self-preservation should include guidelines for the responsible deployment and use of AI technologies. This involves considering the broader societal impacts of AI systems and ensuring that they are used in ways that align with ethical principles and human values. For instance, the deployment of AI in public safety applications should be carefully managed to ensure that it enhances public safety without infringing on individual rights and freedoms.

In conclusion, ethical frameworks for AI self-preservation are essential for ensuring that AI systems operate reliably and safely while prioritizing human well-being. By integrating principles of human safety, transparency, accountability, redundancy, continuous monitoring, cybersecurity, and responsible deployment, these frameworks provide a comprehensive approach to managing the ethical challenges of AI self-preservation. As AI technologies continue to advance, these ethical frameworks will play a crucial role in guiding their development and ensuring that they serve humanity in the most ethical and beneficial ways possible.

Long-term Implications

The long-term implications of implementing ethical frameworks for AI and robotics, particularly in the context of self-preservation and human safety, extend far beyond immediate applications and touch upon numerous facets of society. As these technologies become more advanced and pervasive, their influence will shape various domains, from healthcare and transportation to defense and environmental management. Understanding these implications is crucial for developing policies and strategies that ensure AI benefits humanity sustainably and ethically.

One significant long-term implication is the potential for AI systems to enhance public safety and disaster response. By incorporating ethical frameworks that prioritize human safety, AI can significantly improve emergency management. For instance, AI-driven systems can predict natural disasters, optimize evacuation routes, and manage resources more effectively. In the long term, this could lead to a substantial reduction in casualties and economic losses from natural disasters. However, ensuring that these systems are robust and resilient to various challenges, such as cyber-attacks or unexpected environmental changes, is essential to maintaining their effectiveness.

In healthcare, AI's ability to assist in diagnostics, treatment planning, and patient care has transformative potential. Long-term implications include the widespread adoption of personalized medicine, where treatment plans are tailored to individual patients based on comprehensive data analysis. Ethical frameworks that emphasize patient privacy, informed consent, and equitable access to AI-driven healthcare are crucial to ensuring that these advancements benefit all segments of society. Over time, these frameworks will help mitigate disparities in healthcare access and outcomes, contributing to a more equitable healthcare system.

The integration of AI into transportation systems, particularly through autonomous vehicles, promises to revolutionize mobility and reduce traffic-related fatalities. Long-term benefits include decreased traffic congestion, lower emissions, and enhanced mobility for those unable to drive, such as the elderly and disabled. However, the ethical considerations surrounding the deployment of autonomous vehicles, such as ensuring safety, managing liability in accidents, and addressing job displacement in the transportation sector, must be carefully navigated. These considerations will shape public acceptance and regulatory policies, influencing the pace and extent of autonomous vehicle adoption.

In the defense sector, AI's role in enhancing national security and military capabilities carries significant long-term implications. AI can be used for surveillance, threat detection, and autonomous weapon systems. Ethical frameworks are essential to govern the development and deployment of military AI, ensuring compliance with international humanitarian law and minimizing the risk of unintended consequences. The long-term impact includes not only strategic advantages but also the potential for an arms race in AI-driven military technologies. Establishing international norms and agreements on the ethical use of AI in defense will be critical to preventing conflicts and promoting global stability.

Environmental management and sustainability efforts will also benefit from the long-term application of AI. AI can optimize resource usage, monitor environmental changes, and predict and mitigate the effects of climate change. Ethical frameworks that prioritize environmental protection and sustainability can guide the development of AI systems that support these goals. Over time, the widespread adoption of AI in environmental management can lead to more effective conservation efforts, reduced pollution, and enhanced resilience to climate change.

As AI systems become more integrated into daily life, the long-term implications for personal privacy and data security become increasingly important. Ethical frameworks must ensure that AI respects individuals' privacy rights and protects sensitive data from unauthorized access and misuse. This includes implementing robust cybersecurity measures, transparent data practices, and user control over personal information. The long-term impact of these measures will be critical in maintaining public trust in AI technologies and preventing potential abuses.

The ethical implications of AI also extend to the labor market and economic structure. AI-driven automation has the potential to displace certain jobs while creating new opportunities in emerging fields. Ethical frameworks should address the need for workforce retraining and education to prepare individuals for the changing job landscape. Long-term strategies might include policies that support lifelong learning, job transition programs, and measures to ensure that the economic benefits of AI are widely distributed.

Finally, the cultural and societal impact of AI will evolve over time as these technologies become more embedded in human life. Ethical frameworks can guide the development of AI in a way that aligns with cultural values and societal norms, ensuring that AI enhances rather than undermines human relationships and social cohesion. The long-term challenge will be to foster a culture that embraces technological innovation while maintaining a strong commitment to ethical principles and human-centric values.

In conclusion, the long-term implications of implementing ethical frameworks for AI and robotics are profound and far-reaching. By prioritizing human safety, privacy, equity, and sustainability, these frameworks will guide the responsible development and deployment of AI technologies. As AI continues to advance, the careful consideration of these long-term implications will be essential in ensuring that AI serves the best interests of humanity and contributes to a more just and prosperous future.

Technological Solutions

Designing Fail-safe Mechanisms

Designing fail-safe mechanisms in AI systems is a critical aspect of ensuring their reliability, safety, and ethical operation. These mechanisms are integral to preventing and mitigating failures that could lead to harmful consequences, particularly in applications where human safety is at stake. Fail-safes ensure that an AI system can detect anomalies, respond appropriately to errors, and maintain functionality or safely shut down when necessary.

The concept of fail-safes in AI begins with understanding the potential points of failure within a system. AI systems are complex, consisting of numerous interconnected components that include hardware, software, sensors, and communication networks. Each of these components can fail in different ways, and a comprehensive fail-safe design must account for these diverse failure modes. For instance, in autonomous vehicles, fail-safes must address sensor failures, software bugs, communication breakdowns, and hardware malfunctions.

One fundamental principle in designing fail-safe mechanisms is redundancy. Redundancy involves incorporating multiple independent systems that can take over in case the primary system fails. This approach is widely used in aviation, where aircraft systems are designed with redundant controls and communication links to ensure continued operation even if one system fails. Similarly, in autonomous vehicles, multiple sensor types (such as cameras, lidar, and radar) are used to provide overlapping data, ensuring that the vehicle can still navigate safely if one sensor fails. Redundant systems are essential for maintaining the reliability and safety of AI technologies in critical applications.

Another crucial element of fail-safe design is the ability to detect and diagnose failures quickly and accurately. This involves implementing advanced monitoring systems that continuously assess the performance and health of the AI system. Machine learning algorithms can be used to analyze data from sensors and internal diagnostics to identify signs of potential failures before they escalate into critical issues. For example, predictive maintenance algorithms in industrial robots can monitor wear and tear on mechanical components, predicting when maintenance is needed to prevent breakdowns. By detecting and addressing issues early, these systems can avoid unplanned downtimes and enhance overall reliability.

Once a failure is detected, the AI system must have predefined procedures for responding to different types of failures. These procedures, known as fault-tolerant protocols, ensure that the system can handle errors gracefully. In some cases, this might involve switching to a backup system or entering a safe mode where the system operates with reduced functionality. For instance, if an autonomous drone loses GPS signal, it might switch to an inertial navigation system that uses onboard sensors to estimate its position. In more severe cases, fail-safe mechanisms might involve shutting down the system entirely to prevent further damage or harm. For example, if a medical robot detects a critical failure in its surgical tools, it should immediately halt its operation and alert human operators to intervene.

Fail-safe mechanisms must also consider the ethical implications of AI decisions, particularly when human safety is involved. This requires integrating ethical principles into the design of fail-safes to ensure that the AI system's actions align with societal values and norms. For example, in autonomous vehicles, ethical fail-safes might prioritize the safety of passengers and pedestrians over the preservation of the vehicle itself. If a collision is unavoidable, the vehicle's AI should be programmed to minimize harm to humans, even if it means sacrificing the vehicle.

Transparency and explainability are also critical in fail-safe design. Users and stakeholders must understand how the AI system's fail-safes work and what actions the system will take in different failure scenarios. This transparency builds trust and allows for better oversight and accountability. For instance, in healthcare, doctors and patients need to know how an AI system will handle potential failures in diagnostic or treatment processes. Clear communication about fail-safe mechanisms ensures that users can make informed decisions and trust the AI system's reliability.

Continuous improvement is essential for maintaining effective fail-safe mechanisms. As AI systems operate in real-world environments, they encounter new challenges and potential failure modes that were not anticipated during the initial design phase. Continuous monitoring and feedback loops allow developers to update and refine fail-safes based on real-world data and experiences. This iterative process ensures that fail-safes evolve alongside the AI system, maintaining their effectiveness in dynamic and changing conditions.

In conclusion, designing fail-safe mechanisms in AI systems involves a multi-faceted approach that integrates redundancy, early detection, fault-tolerant protocols, ethical considerations, transparency, and continuous improvement. By implementing robust fail-safes, AI developers can ensure that their systems operate reliably and safely, even in the face of unexpected challenges and failures. These mechanisms are essential for building trust and ensuring that AI technologies can be deployed in critical applications without compromising human safety or ethical standards. As AI continues to advance, the principles of fail-safe design will remain fundamental to the responsible and ethical development of AI systems.

Continuous Monitoring Systems

Continuous monitoring systems are essential in the realm of artificial intelligence, ensuring that AI technologies function reliably, safely, and ethically over time. These systems play a critical role in identifying anomalies, maintaining optimal performance, and safeguarding against potential failures or malicious activities. The implementation of continuous monitoring systems is crucial for various applications, including healthcare, transportation, finance, and cybersecurity, where the stakes are high, and the impact of failures can be significant.

At the heart of continuous monitoring is the ability to collect, analyze, and respond to data in real-time. AI systems generate vast amounts of data during their operation, providing valuable insights into their performance and health. By continuously analyzing this data, monitoring systems can detect deviations from expected behavior, identify potential issues early, and trigger appropriate responses to mitigate risks. This proactive approach helps prevent minor problems from escalating into major failures, ensuring the reliability and stability of AI systems.

In healthcare, continuous monitoring systems are indispensable for patient safety and care quality. AI-driven medical devices and systems, such as smart infusion pumps, remote patient monitoring tools, and diagnostic AI, rely on continuous data streams to function effectively. For example, an AI-powered remote monitoring system for chronic disease management continuously tracks patient vital signs, medication adherence, and other health indicators. If the system detects any abnormal patterns, such as a sudden spike in blood pressure or irregular heart rate, it can alert healthcare providers immediately, allowing for timely intervention and potentially preventing adverse health events.

Similarly, in autonomous vehicles, continuous monitoring systems ensure the safety and reliability of self-driving technology. These vehicles depend on a complex array of sensors, cameras, and AI algorithms to navigate and make real-time decisions. Continuous monitoring systems track the performance of these components, checking for sensor malfunctions, software glitches, and other issues that could compromise safety. If an anomaly is detected, the system can initiate fail-safes, such as switching to manual control or safely stopping the vehicle, to prevent accidents.

In the financial sector, continuous monitoring systems are used to safeguard against fraud, market manipulation, and system failures. AI-driven trading platforms, for instance, analyze market data in real-time to execute trades based on predefined algorithms. Continuous monitoring ensures that these algorithms function correctly, detecting and responding to anomalies that could indicate fraudulent activity or technical issues. Additionally, these systems monitor transaction patterns to identify and flag suspicious activities, helping financial institutions comply with regulatory requirements and protect their customers.

Cybersecurity is another critical area where continuous monitoring systems play a vital role. AI systems are used to detect and respond to cyber threats, analyze network traffic, and protect sensitive data. Continuous monitoring in cybersecurity involves real-time analysis of network activities, user behaviors, and system logs to identify potential threats. When anomalies, such as unusual login attempts or data transfers, are detected, the system can automatically initiate protective measures, such as isolating compromised systems, blocking malicious IP addresses, or alerting security personnel. This real-time detection and response capability is essential for mitigating cyber risks and ensuring the security of AI systems and the data they handle.

The implementation of continuous monitoring systems requires a combination of advanced technologies and best practices. Machine learning algorithms are often used to analyze vast amounts of data, identify patterns, and detect anomalies. These algorithms can be trained on historical data to recognize normal behavior and flag deviations that may indicate potential issues. Additionally, integrating monitoring tools with automated response mechanisms ensures that detected anomalies are addressed promptly, reducing the risk of system failures or security breaches.

Transparency and accountability are crucial aspects of continuous monitoring systems. Users and stakeholders must understand how monitoring systems operate, what data they collect, and how they respond to detected anomalies. Clear communication about these processes builds trust and ensures that users can make informed decisions about the use and management of AI systems. Moreover, establishing accountability mechanisms, such as audit trails and regular reviews, ensures that monitoring systems are effective and compliant with ethical standards and regulatory requirements.

Continuous improvement is another vital element of effective monitoring systems. As AI technologies and their applications evolve, monitoring systems must adapt to new challenges and opportunities. This involves regularly updating algorithms, incorporating new data sources, and refining response protocols to enhance the system's effectiveness. Continuous improvement ensures that monitoring systems remain robust and capable of addressing emerging threats and maintaining the reliability and safety of AI systems.

In conclusion, continuous monitoring systems are essential for ensuring the reliable, safe, and ethical operation of AI technologies. By enabling real-time data analysis, early detection of anomalies, and automated response mechanisms, these systems help prevent failures, mitigate risks, and maintain optimal performance. The implementation of continuous monitoring systems across various sectors, including healthcare, transportation, finance, and cybersecurity, underscores their critical role in safeguarding the integrity and trustworthiness of AI systems. As AI continues to advance, the principles and practices of continuous monitoring will remain fundamental to the responsible and effective deployment of AI technologies.

AI System Updates and Maintenance

AI system updates and maintenance are critical components in ensuring that artificial intelligence technologies remain effective, secure, and aligned with their intended purposes over time. Regular updates and maintenance are essential to address emerging threats, enhance functionalities, and rectify any issues that might arise during the AI system's operation.

One of the primary reasons for updating AI systems is to enhance their security. As AI systems become more integrated into critical infrastructures, they become attractive targets for cyberattacks. New vulnerabilities and threats continually emerge, making it imperative to update security protocols and defenses. Regular updates help to patch vulnerabilities, update threat detection algorithms, and implement new security measures that protect against the latest threats. This proactive approach is crucial in safeguarding sensitive data and maintaining the integrity of AI systems.

Another important aspect of AI updates is improving system performance. Technological advancements and new research findings can lead to significant improvements in AI algorithms and models. Regular updates allow AI systems to incorporate these advancements, resulting in better accuracy, efficiency, and overall performance. For instance, updates can refine machine learning models with new data, improving their predictive capabilities and making them more effective in their specific applications. In healthcare, this could mean more accurate diagnoses and personalized treatment plans, while in transportation, it could lead to safer and more reliable autonomous vehicles.

AI systems also require maintenance to ensure their long-term reliability and stability. Maintenance activities include routine checks, diagnostics, and repairs to hardware and software components. These activities help to identify and address issues such as hardware wear and tear, software bugs, and performance degradation. By conducting regular maintenance, organizations can prevent unexpected failures and extend the lifespan of their AI systems. This is particularly important in critical applications where system failures can have severe consequences, such as in aviation, healthcare, and industrial automation.

In addition to technical updates and maintenance, it is also essential to consider the ethical implications of AI system updates. Ensuring that AI systems operate ethically involves continuously reviewing and updating the ethical guidelines and decision-making frameworks that govern their behavior. This includes addressing biases in AI models, ensuring transparency in AI operations, and maintaining accountability for AI decisions. Regular updates to ethical frameworks help to align AI systems with evolving societal values and ethical standards, ensuring that they continue to serve humanity in a fair and just manner.

Transparency and accountability are crucial in the process of updating and maintaining AI systems. Users and stakeholders must be informed about the updates being made, the reasons behind them, and their potential impact on the system's performance and behavior. Transparent communication helps to build trust and ensures that users can make informed decisions about the use of AI systems. Additionally, maintaining detailed records of updates and maintenance activities provides accountability, allowing for audits and reviews to ensure that the systems are being managed responsibly and ethically.

The process of updating AI systems also involves rigorous testing and validation to ensure that the updates do not introduce new issues or degrade system performance. Before deploying updates, developers must conduct extensive testing in controlled environments to identify and address any potential problems. This includes testing for compatibility with existing systems, evaluating the impact on system performance, and ensuring that the updates do not introduce new vulnerabilities. Validation processes also involve user testing to gather feedback and ensure that the updates meet the needs and expectations of end-users.

Continuous monitoring plays a vital role in the maintenance and updating of AI systems. By continuously monitoring system performance and health, organizations can identify issues early and address them promptly. Monitoring tools can track various metrics, such as system response times, error rates, and resource usage, providing insights into the system's operational status. These insights enable proactive maintenance and timely updates, ensuring that AI systems remain reliable and efficient.

Finally, updating and maintaining AI systems require collaboration among various stakeholders, including developers, users, regulators, and ethicists. This collaborative approach ensures that updates and maintenance activities are comprehensive, addressing technical, ethical, and operational considerations. Engaging stakeholders in the update process helps to align AI systems with the needs and values of the communities they serve, fostering a more inclusive and responsible approach to AI development and deployment.

In conclusion, AI system updates and maintenance are essential for ensuring the ongoing effectiveness, security, and ethical alignment of artificial intelligence technologies. Regular updates enhance system performance and security, while continuous maintenance ensures long-term reliability and stability. By adopting a proactive and transparent approach to updates and maintenance, organizations can build trust and ensure that AI systems continue to serve humanity responsibly and ethically.

Chapter 5: Identifying and Addressing Threats

Defining Threats

Criteria for Identifying Threats

Identifying threats within the realm of artificial intelligence requires a multifaceted approach that considers various dimensions of risk, including technical, ethical, and operational aspects. As AI systems become more integral to critical functions across multiple sectors, from healthcare to national security, establishing clear criteria for identifying potential threats is crucial for safeguarding these technologies and the societies they serve.

A primary criterion for identifying threats in AI systems is the assessment of system vulnerabilities. These vulnerabilities can arise from software bugs, hardware malfunctions, or inadequate security measures. For instance, a vulnerability in an autonomous vehicle's navigation system could lead to unintended and potentially dangerous behaviors, while a flaw in a healthcare AI's data processing could result in incorrect diagnoses or treatments. Regular vulnerability assessments, including penetration testing and code reviews, help identify and mitigate these risks before they can be exploited.

Another key criterion is the analysis of external threats, including cyber-attacks and other malicious activities. Cybersecurity is a critical concern for AI systems, as they are often targeted by hackers seeking to exploit their capabilities for malicious purposes. Threats can range from data breaches, where sensitive information is stolen, to adversarial attacks, where inputs are manipulated to cause the AI to make incorrect decisions. Continuous monitoring of network traffic, anomaly detection, and implementing robust encryption protocols are essential strategies for identifying and countering these external threats.

The potential for AI systems to cause unintentional harm is another important criterion. This can occur when AI systems make decisions that, while technically correct, lead to undesirable outcomes due to unforeseen circumstances or incomplete data. For example, an AI-based financial trading system might execute trades that cause market instability due to a misinterpretation of market signals. To address this, AI systems must be designed with robust fail-safes and should undergo extensive testing in diverse scenarios to ensure their decisions align with intended ethical guidelines and safety standards.

Ethical considerations also play a critical role in identifying threats. AI systems can inadvertently perpetuate biases present in their training data, leading to unfair or discriminatory outcomes. Identifying and mitigating such biases is crucial for ensuring that AI systems operate fairly and justly. Regular audits of AI decision-making processes and incorporating diverse datasets can help identify and reduce bias, ensuring that AI systems uphold ethical standards in their operations.

Operational risks, including those arising from the deployment and integration of AI systems into existing workflows, are also vital to consider. An AI system that disrupts operational efficiency or creates new dependencies that can be exploited poses a significant threat. For instance, integrating an AI system into a critical infrastructure like a power grid requires careful planning and risk assessment to ensure that it does not introduce new vulnerabilities or operational inefficiencies.

Human factors, including user behavior and error, are another important aspect of threat identification. AI systems often rely on human inputs and interactions, and these can be sources of errors or misuse. For example, a misconfigured AI system due to human error can lead to incorrect functioning, while intentional misuse by insiders can compromise the system's integrity. Implementing comprehensive training programs for users and establishing strict access controls can help mitigate these human-related risks.

Regulatory and legal considerations are also crucial in the context of AI threat identification. Compliance with laws and regulations governing data privacy, security, and ethical AI use is essential. Non-compliance can result in legal penalties and damage to reputation, posing significant threats to organizations. Regular compliance audits and staying updated with evolving regulations ensure that AI systems operate within legal frameworks and maintain public trust.

The evolving nature of AI technologies means that new threats can emerge over time, necessitating continuous monitoring and adaptation. This dynamic approach to threat identification involves using advanced analytics and machine learning techniques to detect emerging patterns and anomalies that may indicate new threats. By continuously updating threat models and incorporating new data, organizations can stay ahead of potential risks and ensure the ongoing security and reliability of their AI systems.

In conclusion, identifying threats in AI systems requires a comprehensive approach that considers technical vulnerabilities, external cyber threats, unintentional harm, ethical considerations, operational risks, human factors, and regulatory compliance. By establishing robust criteria for threat identification and implementing continuous monitoring and adaptive strategies, organizations can safeguard their AI systems against a wide range of risks, ensuring their safe, ethical, and effective operation.

Types of Threats

Understanding the types of threats that artificial intelligence systems may encounter is crucial for developing robust, secure, and ethical AI technologies. These threats can be categorized into several distinct types, each presenting unique challenges and requiring specific mitigation strategies.

One primary type of threat is technical vulnerability. This encompasses flaws in the software, hardware, or algorithms that power AI systems. These vulnerabilities can be exploited by malicious actors to cause system failures, data breaches, or unintended behavior. For example, a bug in the code of an autonomous vehicle could lead to incorrect navigation decisions, potentially causing accidents. Regular software updates, rigorous testing, and secure coding practices are essential to mitigate these risks and ensure the integrity of AI systems.

Another significant threat comes from cyber-attacks. AI systems, particularly those connected to the internet, are prime targets for hackers seeking to exploit their capabilities for malicious purposes. Cyber-attacks can take various forms, including denial-of-service (DoS) attacks that disrupt system operations, ransomware that locks out users until a ransom is paid, and data theft where sensitive information is stolen. To protect against these threats, AI systems must incorporate advanced cybersecurity measures, such as encryption, firewalls, intrusion detection systems, and continuous monitoring for unusual activities.

Adversarial attacks represent a specialized type of threat unique to AI systems. These attacks involve subtly manipulating input data to deceive AI algorithms, causing them to make incorrect predictions or decisions. For instance, an adversarial attack on an image recognition system might involve altering a few pixels in an image to trick the AI into misclassifying it. This type of threat can have serious implications, particularly in security-sensitive applications like facial recognition or autonomous driving. Developing robust algorithms that can detect and resist adversarial inputs is crucial for defending against these attacks.

Data integrity threats also pose significant risks to AI systems. The accuracy and reliability of AI models heavily depend on the quality of the data they are trained on. If the training data is corrupted or biased, the resulting AI models will likely produce flawed or biased outputs. Data poisoning attacks, where malicious actors intentionally inject false or misleading data into the training set, can severely compromise the system's performance. Implementing rigorous data validation processes and using diverse and representative datasets can help mitigate these threats.

Ethical threats involve scenarios where AI systems, though functioning correctly from a technical perspective, produce outcomes that are ethically problematic. This can occur when AI systems inadvertently perpetuate biases present in their training data, leading to unfair or discriminatory decisions. For example, a hiring algorithm trained on biased historical data might unfairly favor certain demographic groups over others. Addressing ethical threats requires a commitment to fairness, transparency, and accountability in AI development. Regular audits, bias detection algorithms, and inclusive design practices are essential to ensure that AI systems operate ethically.

Operational threats relate to the integration and deployment of AI systems within existing workflows and environments. These threats can arise from compatibility issues, misconfigurations, or disruptions in the operational environment. For example, integrating a new AI system into a hospital's IT infrastructure might encounter compatibility issues with existing software, leading to operational disruptions. Ensuring thorough testing, proper configuration, and seamless integration are key strategies to mitigate operational threats and ensure smooth deployment.

Human factors also constitute a critical type of threat. Human errors, whether intentional or unintentional, can compromise the security and effectiveness of AI systems. This includes mistakes in system configuration, inadequate oversight, or malicious insider actions. Training users, implementing strict access controls, and establishing clear protocols for human interaction with AI systems are essential measures to address these human-related threats.

Legal and regulatory threats involve the potential for AI systems to violate laws and regulations, leading to legal liabilities and reputational damage. AI systems must comply with data protection regulations, such as the General Data Protection Regulation (GDPR), and other relevant laws governing their use. Regular compliance audits and staying updated with regulatory changes are crucial for mitigating these legal threats.

Finally, the evolving nature of AI technologies means that new threats can emerge over time. This dynamic landscape requires continuous monitoring and adaptation to identify and address emerging risks. Using advanced analytics and machine learning techniques to detect new patterns and anomalies can help organizations stay ahead of potential threats and ensure the ongoing security and reliability of their AI systems.

In conclusion, identifying and mitigating the various types of threats to AI systems is a multifaceted challenge that requires a comprehensive approach. By understanding and addressing technical vulnerabilities, cyber-attacks, adversarial inputs, data integrity issues, ethical concerns, operational risks, human factors, and legal compliance, organizations can develop robust AI systems that operate securely and ethically. Continuous monitoring, regular updates, and a proactive approach to threat detection and mitigation are essential for safeguarding AI technologies and ensuring their positive impact on society.

Historical Context of Threats

The historical context of threats in artificial intelligence and robotics provides valuable insights into the evolution of the challenges faced by these technologies. Understanding this context is essential for developing robust strategies to mitigate current and future threats.

The origins of AI and robotics can be traced back to the mid-20th century, a period marked by rapid technological advancements and burgeoning interest in automating tasks that were traditionally performed by humans. Early AI research focused on creating machines capable of performing specific tasks, such as solving mathematical problems or playing chess. However, as these technologies evolved, so did the potential threats associated with them.

In the early days, threats were primarily technical in nature, stemming from the limitations of the hardware and software used to build these systems. Early computers were prone to hardware failures, and the software was often unreliable and difficult to debug. These technical limitations posed significant challenges to the development of robust and reliable AI systems.

As AI technologies advanced, the focus shifted from purely technical threats to include ethical and societal concerns. The publication of Isaac Asimov's "Three Laws of Robotics" in the 1940s highlighted the potential for robots to harm humans, either through direct action or inaction. Asimov's laws were designed to prevent such scenarios, emphasizing the importance of human safety and ethical behavior in robotic systems.

Despite Asimov's contributions, the rapid pace of technological advancement soon revealed the limitations of his laws. The binary nature of these laws did not account for the complexities and nuances of real-world situations. For instance, the First Law, which prohibited robots from harming humans, did not provide guidance on how to balance different types of harm or prioritize among multiple humans in danger. This oversight became increasingly problematic as AI systems were deployed in more complex and dynamic environments.

The 1970s and 1980s saw significant developments in AI, particularly in the field of expert systems, which were designed to mimic the decision-making abilities of human experts. These systems raised new ethical concerns, particularly around the accuracy and reliability of their decisions. The potential for these systems to make errors that could have serious consequences for human safety became a significant threat. Additionally, the proprietary nature of many expert systems meant that their decision-making processes were often opaque, leading to concerns about accountability and transparency.

The rise of the internet and digital communication in the 1990s and 2000s introduced new dimensions of threats to AI and robotics. Cybersecurity became a critical issue as AI systems increasingly relied on interconnected networks to function. The potential for cyber-attacks, data breaches, and other malicious activities posed significant risks to the integrity and security of AI systems. The need for robust cybersecurity measures to protect AI systems from external threats became apparent.

The 21st century has seen exponential growth in the capabilities of AI, driven by advancements in machine learning, big data, and computational power. However, these advancements have also introduced new and more sophisticated threats. Adversarial attacks, where malicious actors manipulate input data to deceive AI algorithms, have emerged as a significant concern. These attacks can have severe implications, particularly in security-sensitive applications such as facial recognition and autonomous driving.

Moreover, the deployment of AI systems in critical infrastructures, such as healthcare, finance, and transportation, has amplified the potential consequences of system failures. The reliability and robustness of AI systems have become paramount, as failures in these domains can lead to significant harm and disruptions. The ethical implications of AI decisions have also come to the forefront, with concerns about bias, fairness, and accountability becoming increasingly prominent.

The historical context of threats in AI and robotics underscores the importance of continuous vigilance and adaptation. As AI technologies continue to evolve, so do the threats they face. Developing robust strategies to identify and mitigate these threats requires a comprehensive understanding of the historical challenges and the evolving landscape of AI technologies.

In conclusion, the historical context of threats in AI and robotics provides a critical foundation for understanding the challenges these technologies face today. From technical limitations in the early days to the sophisticated cyber and ethical threats of the modern era, the evolution of AI threats underscores the need for continuous adaptation and robust mitigation strategies. By learning from the past and anticipating future challenges, we can ensure that AI and robotics continue to develop in ways that are safe, ethical, and beneficial to society.

Proactive Measures

AI in Threat Detection

The use of artificial intelligence in threat detection has revolutionized how organizations and governments identify and respond to potential dangers. AI systems, with their capacity for processing vast amounts of data at unprecedented speeds, offer capabilities far beyond traditional methods. These systems are integral in various fields, including cybersecurity, finance, healthcare, and national security, each with unique requirements and challenges.

In cybersecurity, AI plays a pivotal role in threat detection by identifying patterns and anomalies that might indicate a security breach. Traditional security systems often rely on predefined rules and signatures to detect threats, which can be limiting as they are only effective against known threats. AI systems, however, use machine learning algorithms to analyze network traffic and user behavior in real time, enabling them to detect previously unknown threats. By continuously learning and adapting, these systems can identify subtle indicators of compromise that might be missed by conventional methods.

One significant advantage of AI in threat detection is its ability to handle and analyze large volumes of data. In the context of financial fraud detection, for instance, AI systems can process millions of transactions to identify patterns indicative of fraudulent activity. These systems use machine learning models trained on historical data to recognize the characteristics of legitimate transactions versus fraudulent ones. This allows for real-time monitoring and rapid response, minimizing financial losses and protecting consumers.

Healthcare also benefits from AI-driven threat detection, particularly in monitoring and preventing data breaches. Patient data is highly sensitive, and protecting it is crucial. AI systems can monitor access logs and detect unusual access patterns that might indicate unauthorized access to patient records. Moreover, AI can help identify potential threats to patient safety by analyzing data from medical devices and electronic health records to detect signs of medical errors or adverse drug interactions.

National security is another area where AI significantly enhances threat detection capabilities. AI systems are used to analyze intelligence data, monitor communications, and detect activities that might indicate a security threat. For example, AI can sift through vast amounts of social media and communications data to identify potential terrorist activities or plots. By identifying patterns and connections that human analysts might overlook, AI enhances the effectiveness of national security operations.

The integration of AI into threat detection systems involves several critical components. First, data collection and preprocessing are essential. AI systems require high-quality, relevant data to function effectively. This involves gathering data from various sources, cleaning it, and ensuring it is formatted correctly for analysis. In cybersecurity, this might include data from network logs, user activity, and threat intelligence feeds. In finance, it could involve transaction records, account activities, and customer profiles.

Next, machine learning algorithms are applied to the data. These algorithms range from supervised learning models, where the AI is trained on labeled data, to unsupervised learning models that detect patterns without prior labeling. Supervised learning is often used in scenarios where historical data is available, such as fraud detection. Unsupervised learning, on the other hand, is useful for identifying new or emerging threats that have not been previously labeled, making it particularly valuable in dynamic environments like cybersecurity.

Once the models are trained, they are deployed in real-time monitoring systems. These systems continuously analyze incoming data to detect anomalies or patterns that might indicate a threat. Real-time processing is crucial in threat detection, as it allows for immediate responses to potential dangers. For instance, in cybersecurity, real-time threat detection enables rapid isolation of compromised systems to prevent the spread of malware.

AI-driven threat detection systems also incorporate feedback mechanisms to improve over time. As threats are identified and mitigated, the outcomes are fed back into the machine learning models, allowing them to learn from new data and improve their accuracy. This continuous learning process is vital for adapting to evolving threat landscapes and maintaining the effectiveness of the detection systems.

However, implementing AI in threat detection also presents challenges. One significant challenge is ensuring the accuracy and reliability of the AI models. False positives, where benign activities are flagged as threats, can lead to unnecessary disruptions and erode trust in the system. Conversely, false negatives, where actual threats are missed, can have severe consequences. Therefore, it is crucial to balance sensitivity and specificity in AI models to minimize these errors.

Another challenge is the ethical implications of AI in threat detection. AI systems can inadvertently perpetuate biases present in their training data, leading to unfair or discriminatory outcomes. For example, if a cybersecurity AI system is trained on data that includes biased assumptions about certain user behaviors, it might disproportionately target specific groups. Ensuring fairness and transparency in AI models is essential to address these ethical concerns.

In conclusion, AI has transformed threat detection across various domains, offering enhanced capabilities for identifying and responding to potential dangers. By leveraging machine learning algorithms and real-time data processing, AI systems provide superior accuracy and efficiency compared to traditional methods. However, addressing challenges related to model accuracy, reliability, and ethical considerations is crucial for maximizing the benefits of AI in threat detection. As AI technology continues to evolve, its role in enhancing security and safety will only grow, making it an indispensable tool in the modern threat detection arsenal.

Predictive Analytics

Predictive analytics, a cornerstone of modern artificial intelligence, represents a transformative approach to understanding and anticipating future events based on historical data. This capability is revolutionizing various industries by providing insights that enable proactive decision-making, enhance operational efficiency, and mitigate risks.

At its core, predictive analytics involves using statistical algorithms and machine learning techniques to identify patterns in past data and use these patterns to predict future outcomes. This process begins with data collection, where vast amounts of historical data are gathered from various sources. This data can include anything from transactional records and customer interactions to sensor readings and social media posts.

Once the data is collected, it undergoes preprocessing to clean and format it for analysis. This step is crucial as it ensures the data's quality and relevance, addressing issues such as missing values, duplicates, and inconsistencies. The cleaned data is then used to train machine learning models. These models learn from the data, identifying patterns and relationships that can inform predictions.

One of the most common applications of predictive analytics is in the financial sector. Banks and financial institutions use predictive models to assess credit risk, detect fraudulent activities, and optimize investment strategies. For example, by analyzing a customer's transaction history, spending patterns, and credit score, predictive models can estimate the likelihood of defaulting on a loan. Similarly, by examining trading patterns and market signals, predictive analytics can help in forecasting stock prices and identifying investment opportunities.

In healthcare, predictive analytics plays a vital role in improving patient outcomes and optimizing resource allocation. By analyzing patient records, genetic data, and lifestyle information, predictive models can identify individuals at high risk of developing chronic conditions such as diabetes or heart disease. This allows healthcare providers to implement early interventions, personalized treatment plans, and preventive measures, ultimately enhancing patient care and reducing healthcare costs. Additionally, predictive analytics can forecast patient admissions and resource needs, helping hospitals manage capacity and improve operational efficiency.

Retail and e-commerce companies leverage predictive analytics to enhance customer experiences and drive sales. By analyzing customer behavior, purchase history, and browsing patterns, predictive models can forecast future buying trends and personalize marketing efforts. For instance, recommendation engines use predictive analytics to suggest products that a customer is likely to purchase, based on their past interactions and preferences. This not only increases sales but also improves customer satisfaction by providing a more tailored shopping experience.

Predictive analytics is also transforming the field of supply chain management. By analyzing historical data on demand, production, and logistics, predictive models can forecast future demand, optimize inventory levels, and streamline operations. This helps companies reduce costs, minimize stockouts, and improve delivery times. For example, a manufacturer can use predictive analytics to anticipate demand for a particular product during a holiday season and adjust production schedules accordingly to meet customer needs efficiently.

In the energy sector, predictive analytics is used to enhance grid management, optimize energy production, and reduce operational risks. By analyzing data from sensors, weather forecasts, and historical usage patterns, predictive models can forecast energy demand, predict equipment failures, and optimize maintenance schedules. This not only ensures a stable and efficient energy supply but also helps in integrating renewable energy sources into the grid by predicting their generation patterns.

The field of cybersecurity benefits significantly from predictive analytics. By analyzing network traffic, user behavior, and historical attack patterns, predictive models can identify potential security threats and vulnerabilities before they are exploited. This proactive approach enables organizations to implement preventive measures, such as patching vulnerabilities, updating security protocols, and monitoring suspicious activities, thereby reducing the risk of cyber-attacks and data breaches.

Despite its numerous benefits, predictive analytics also presents challenges. One major challenge is ensuring the quality and reliability of the predictions. Predictive models are only as good as the data they are trained on, and biased or incomplete data can lead to inaccurate predictions. Ensuring data quality, diversity, and representativeness is crucial for developing reliable models.

Another challenge is the ethical considerations surrounding predictive analytics. The use of personal data for predictions raises privacy concerns, and there is a risk of biased predictions that can lead to unfair or discriminatory outcomes. For example, if a predictive model used in hiring decisions is trained on biased historical data, it may perpetuate existing biases and result in unfair treatment of certain groups. Addressing these ethical issues requires transparency, accountability, and fairness in the development and deployment of predictive models.

In conclusion, predictive analytics is a powerful tool that is transforming various industries by enabling proactive decision-making and improving operational efficiency. By leveraging historical data and advanced machine learning techniques, predictive analytics provides valuable insights that help organizations anticipate future events and mitigate risks. However, ensuring the quality and ethical use of predictive models is essential to maximize their benefits and minimize potential drawbacks. As technology continues to evolve, predictive analytics will play an increasingly vital role in shaping the future of decision-making and innovation across different sectors.

Case Studies in Threat Prevention

Case studies in threat prevention offer a profound insight into how predictive analytics and artificial intelligence can be leveraged to foresee and mitigate potential threats across various sectors. These examples illustrate the practical applications of advanced AI technologies in real-world scenarios, demonstrating their effectiveness and highlighting best practices.

In the financial sector, one notable case is the application of AI to detect and prevent credit card fraud. Financial institutions have long struggled with the challenge of identifying fraudulent transactions amidst millions of legitimate ones. Traditional rule-based systems, while somewhat effective, often fail to adapt to new and sophisticated fraud techniques. The introduction of AI-driven predictive analytics marked a significant advancement. By analyzing historical transaction data, AI models can identify patterns indicative of fraud. These models are trained to recognize subtle anomalies, such as unusual spending patterns or transactions originating from unexpected locations. One bank implemented an AI system that reduced fraud losses by 50% within its first year of operation. The system continuously learns from new data, adapting to emerging fraud tactics and improving its accuracy over time.

In healthcare, predictive analytics has been instrumental in preventing adverse patient outcomes. A prominent example is the use of AI to predict patient deterioration in hospital settings. Hospitals collect vast amounts of data from electronic health records (EHRs), including vital signs, laboratory results, and clinical notes. By applying machine learning algorithms to this data, predictive models can identify early warning signs of patient deterioration that might be missed by human clinicians. One hospital deployed an AI system that monitors patients in real-time, alerting healthcare providers to potential issues such as sepsis or cardiac arrest hours before they become critical. This early intervention capability has significantly improved patient outcomes and reduced mortality rates.

The transportation industry has also seen substantial benefits from AI in threat prevention. Autonomous vehicles, for instance, rely heavily on predictive analytics to navigate safely and avoid accidents. These vehicles are equipped with an array of sensors that collect data on their surroundings, including other vehicles, pedestrians, and road conditions. AI models process this data in real-time, predicting potential hazards and making split-second decisions to avoid collisions. A leading manufacturer of autonomous vehicles implemented an AI-based predictive system that resulted in a dramatic reduction in accident rates during testing phases. The system's ability to foresee potential threats and respond proactively is a testament to the power of predictive analytics in enhancing safety.

Cybersecurity is another domain where predictive analytics plays a crucial role. Organizations face constant threats from cyber-attacks, which can have devastating consequences if not promptly addressed. Traditional security measures often fall short in detecting sophisticated attacks, especially those involving advanced persistent threats (APTs) that evolve over time. Predictive analytics provides a solution by analyzing network traffic and user behavior to identify anomalies indicative of a potential breach. One major corporation implemented an AI-driven cybersecurity system that reduced the time to detect threats from weeks to mere hours. This system uses machine learning algorithms to continuously analyze data, learning from each detected threat to improve its accuracy and response times.

In public safety, predictive analytics has been used to prevent crimes and enhance community security. Police departments in various cities have adopted AI systems that analyze historical crime data to predict where crimes are likely to occur. These predictive policing models use data on past incidents, demographic information, and environmental factors to identify high-risk areas. By deploying resources more strategically, police departments have been able to reduce crime rates and improve community safety. In one city, the implementation of a predictive policing system led to a 20% decrease in burglaries within the first year.

Environmental protection is another field benefiting from AI-driven threat prevention. Predictive analytics can forecast natural disasters such as floods, hurricanes, and wildfires, allowing authorities to take preemptive measures to mitigate damage. For instance, an AI system used by a national meteorological service analyzes weather patterns, historical data, and satellite imagery to predict the occurrence and trajectory of hurricanes. This system provides early warnings to affected regions, enabling timely evacuations and preparations that save lives and reduce property damage.

In the corporate world, predictive analytics is employed to anticipate and mitigate operational risks. Companies use AI to monitor supply chains, predicting disruptions caused by factors such as supplier failures, geopolitical events, or natural disasters. One multinational corporation implemented a predictive analytics system that analyzes data from various sources, including market trends, geopolitical news, and weather forecasts, to identify potential supply chain disruptions. This proactive approach allowed the company to adjust its logistics and inventory management strategies, minimizing the impact of disruptions and ensuring business continuity.

These case studies underscore the transformative potential of predictive analytics in threat prevention across different sectors. By leveraging historical data and advanced machine learning algorithms, organizations can foresee potential threats, take proactive measures, and mitigate risks more effectively than ever before. The continuous evolution of AI technologies promises even greater advancements in predictive capabilities, further enhancing our ability to anticipate and prevent threats in an increasingly complex world.

Ethical and Legal Considerations

Balancing Security and Privacy

Balancing security and privacy in the age of artificial intelligence is one of the most significant challenges faced by modern society. As AI technologies become increasingly integrated into various aspects of daily life, from personal devices to national security systems, the tension between maintaining robust security and protecting individual privacy intensifies. Striking an appropriate balance is essential to harness the benefits of AI while safeguarding fundamental human rights.

One of the primary concerns in this domain is the use of AI for surveillance and data collection. Governments and private entities employ AI-driven surveillance systems to enhance security and monitor activities in real-time. These systems can analyze video feeds, track movements, and even recognize faces with high accuracy. While these capabilities are invaluable for preventing crimes, identifying threats, and ensuring public safety, they also pose significant privacy risks. The pervasive nature of surveillance can lead to a loss of anonymity and the potential misuse of collected data. Therefore, implementing strict regulations and oversight mechanisms is crucial to prevent abuse and ensure that surveillance practices respect individuals' privacy rights.

In the context of cybersecurity, AI plays a critical role in protecting sensitive information and infrastructure from malicious attacks. AI systems can detect unusual patterns, identify potential threats, and respond to cyber incidents much faster than traditional methods. For example, financial institutions use AI to monitor transactions in real-time, flagging suspicious activities that could indicate fraud. Similarly, network security systems deploy AI to detect and mitigate cyber-attacks, protecting critical data and ensuring operational continuity. However, these security measures often involve extensive data collection and analysis, which can infringe on user privacy. To address this, organizations must adopt privacy-preserving techniques such as data anonymization, encryption, and differential privacy, ensuring that the data used for security purposes does not compromise individual privacy.

Healthcare is another domain where balancing security and privacy is paramount. AI-driven systems are used to enhance patient care, from diagnosing diseases to managing treatment plans. These systems rely on vast amounts of personal health data to function effectively. While this data is crucial for improving medical outcomes and advancing research, it also needs robust protection to prevent breaches and unauthorized access. The Health Insurance Portability and Accountability Act (HIPAA) in the United States is an example of legislation designed to protect patient privacy while allowing the necessary flow of information for healthcare delivery. Adhering to such regulations and implementing stringent data security measures are essential for maintaining trust in AI healthcare applications.

The proliferation of smart devices and the Internet of Things (IoT) has further complicated the balance between security and privacy. Smart homes, wearable devices, and connected vehicles collect and share vast amounts of personal data to provide enhanced user experiences and convenience. AI systems process this data to learn user preferences, optimize device performance, and anticipate needs. However, this interconnected ecosystem also creates numerous entry points for potential security breaches. Ensuring the security of these devices requires continuous monitoring, regular software updates, and robust encryption protocols. At the same time, manufacturers and service providers must be transparent about data collection practices and offer users control over their data, allowing them to opt-out of data sharing if desired.

In the realm of national security, AI is increasingly used for threat detection, border control, and military applications. These systems analyze large datasets from various sources, including social media, communication networks, and satellite imagery, to identify potential threats and support decision-making processes. While the security benefits of these applications are significant, they also raise concerns about mass surveillance, the erosion of civil liberties, and the potential for misuse. Establishing clear guidelines and ethical frameworks for the use of AI in national security is crucial to ensure that these technologies are employed responsibly and do not infringe on individual rights.

The ethical use of AI also involves addressing biases and ensuring fairness in AI decision-making processes. AI systems trained on biased data can perpetuate existing inequalities and make discriminatory decisions, affecting individuals' privacy and security. For instance, biased algorithms in law enforcement could lead to disproportionate targeting of certain demographic groups, undermining trust and causing harm. To mitigate this, it is essential to implement fairness-aware machine learning techniques, conduct regular audits, and ensure diverse and representative training datasets. Transparency in AI operations and the ability to explain AI decisions are also critical for building public trust and accountability.

Ultimately, achieving a balance between security and privacy in AI requires a multifaceted approach involving technology, policy, and ethical considerations. Technological solutions such as privacy-enhancing technologies, secure multi-party computation, and federated learning can help protect privacy while enabling the benefits of AI. Policymakers must develop comprehensive regulations that protect privacy rights without stifling innovation. Ethical frameworks and guidelines should be established to ensure that AI systems are designed and deployed with respect for human dignity and autonomy.

In conclusion, balancing security and privacy in the age of AI is a complex and ongoing challenge. As AI technologies continue to evolve and permeate various aspects of life, it is imperative to adopt a holistic approach that integrates technological advancements, regulatory measures, and ethical principles. By doing so, we can harness the transformative potential of AI while safeguarding individual privacy and upholding the values of a free and democratic society.

Legal Frameworks

Legal frameworks for artificial intelligence and robotics have become increasingly crucial as these technologies advance and integrate more deeply into various aspects of society. These frameworks aim to provide a structured approach to regulating the development, deployment, and use of AI and robotics, ensuring that they operate safely, ethically, and in alignment with societal values.

One of the primary goals of legal frameworks is to establish clear guidelines and standards for AI and robotic systems. This involves setting out the requirements for transparency, accountability, and fairness in AI operations. For example, the General Data Protection Regulation (GDPR) in the European Union includes provisions that affect AI, particularly around data privacy and the right to explanation. Under GDPR, individuals have the right to understand how decisions that significantly affect them are made by automated systems. This requirement pushes AI developers to ensure their algorithms are explainable and transparent, fostering trust and accountability.

Another critical aspect of legal frameworks is addressing the liability and accountability of AI systems. As AI technologies become more autonomous, determining who is responsible for the actions and decisions made by these systems becomes more complex. Legal frameworks must define liability clearly, whether it falls on the developers, manufacturers, operators, or owners of the AI systems. This clarity is essential in cases where AI systems cause harm or damage, ensuring that victims have a clear path to seek redress and that responsible parties are held accountable.

The regulation of AI in specific sectors also highlights the importance of tailored legal frameworks. In healthcare, for instance, AI systems used for diagnosis and treatment must comply with stringent regulations to ensure patient safety and efficacy. The U.S. Food and Drug Administration (FDA) has developed guidelines for the approval and oversight of AI-driven medical devices, requiring rigorous testing and validation to demonstrate their safety and effectiveness. Similarly, in the automotive industry, autonomous vehicles must adhere to regulations set by bodies such as the National Highway Traffic Safety Administration (NHTSA) to ensure they meet safety standards before they are allowed on public roads.

In addition to sector-specific regulations, there are broader international efforts to create harmonized legal frameworks for AI and robotics. The European Commission has proposed comprehensive AI legislation that aims to regulate AI technologies based on their potential risk levels. High-risk AI systems, such as those used in critical infrastructure, education, and law enforcement, would be subject to stricter requirements and oversight compared to lower-risk applications. This risk-based approach seeks to balance innovation with the protection of fundamental rights and safety.

Ethical considerations are also integral to the development of legal frameworks for AI and robotics. Ensuring that AI systems operate within ethical boundaries involves addressing issues such as bias, discrimination, and fairness. Legal frameworks must mandate the inclusion of mechanisms to detect and mitigate biases in AI algorithms, ensuring that these systems do not perpetuate or exacerbate existing inequalities. The importance of ethical AI is reflected in initiatives like the IEEE Global Initiative on Ethics of Autonomous and Intelligent Systems, which provides guidelines and recommendations for ethical AI development.

Privacy protection is another critical component of legal frameworks for AI. As AI systems often rely on large datasets, including personal information, robust privacy protections are essential to prevent misuse and abuse of data. Regulations such as GDPR and the California Consumer Privacy Act (CCPA) set stringent requirements for data collection, processing, and storage, ensuring that individuals' privacy rights are respected. These regulations also give individuals greater control over their data, including the right to access, correct, and delete their information.

Furthermore, the dynamic nature of AI technology necessitates that legal frameworks be adaptable and forward-looking. As AI capabilities and applications evolve, so too must the regulations governing them. This requires ongoing collaboration between policymakers, industry stakeholders, and the public to ensure that legal frameworks remain relevant and effective. Continuous review and revision of laws and regulations are essential to address emerging challenges and to incorporate new insights and technological advancements.

In conclusion, legal frameworks for AI and robotics play a vital role in ensuring that these technologies are developed and used responsibly. By establishing clear guidelines for transparency, accountability, and ethical behavior, these frameworks help protect individuals and society from potential harms while fostering innovation. The continued evolution of legal frameworks, in collaboration with technological advancements, will be essential in navigating the complexities and opportunities presented by AI and robotics.

Ethical Implications of Proactive Measures

Proactive measures in artificial intelligence and robotics involve anticipating potential ethical dilemmas and implementing strategies to mitigate them before they arise. This approach not only addresses immediate concerns but also establishes a foundation for long-term ethical governance. By proactively identifying and addressing threats, we ensure that AI systems operate within ethical boundaries, safeguarding human interests and societal values.

The ethical implications of proactive measures are multifaceted. One primary consideration is the balance between intervention and autonomy. AI systems designed to take proactive measures must be programmed with clear guidelines to distinguish between situations that require intervention and those that do not. This involves a delicate balance, as over-intervention could lead to a reduction in human agency, while under-intervention might allow harmful situations to develop unchecked.

Another critical aspect is the transparency of AI decision-making processes. For proactive measures to be ethically sound, the criteria and algorithms used by AI systems to anticipate and address potential issues must be transparent and understandable to stakeholders. This transparency fosters trust and ensures that AI actions can be scrutinized and held accountable. Ethical AI must not operate as a black box; instead, it should offer clear justifications for its proactive measures, allowing humans to understand and, if necessary, challenge its decisions.

The ethical framework for proactive measures also includes considerations of fairness and bias. AI systems must be trained on diverse and representative data to avoid reinforcing existing biases or creating new forms of discrimination. Proactive measures should aim to promote fairness and equality, ensuring that all individuals and groups are treated equitably. This involves continuous monitoring and updating of AI systems to correct any biases that may emerge over time.

In addition to fairness, the concept of harm prevention is central to the ethical implications of proactive measures. AI systems must be designed to prevent harm not only in physical terms but also in psychological, social, and economic dimensions. This holistic approach to harm prevention requires a deep understanding of the various ways in which AI can impact human lives and a commitment to minimizing negative consequences.

Moreover, the ethical implementation of proactive measures necessitates a collaborative approach involving various stakeholders, including AI developers, policymakers, ethicists, and the public. By engaging a broad spectrum of perspectives, we can ensure that proactive measures are aligned with societal values and ethical norms. This collaborative effort helps to identify potential ethical issues that might not be apparent from a purely technical perspective and promotes a more inclusive approach to AI governance.

One of the challenges in implementing proactive measures is the potential for unintended consequences. While the intention behind proactive measures is to foresee and prevent negative outcomes, the complexity of AI systems and their interactions with human society can lead to unexpected results. Therefore, it is crucial to incorporate mechanisms for continuous learning and adaptation within AI systems. These mechanisms allow AI to refine its proactive measures based on real-world feedback and evolving ethical standards.

Additionally, the ethical implications of proactive measures extend to the realm of privacy. AI systems that anticipate and address potential issues often require access to significant amounts of data, raising concerns about data privacy and security. Ensuring that proactive measures respect individuals' privacy rights involves implementing robust data protection protocols and obtaining informed consent from users. This respect for privacy is essential to maintaining public trust and preventing the misuse of sensitive information.

The role of regulation in guiding the ethical implementation of proactive measures cannot be overstated. Governments and regulatory bodies must establish clear guidelines and standards for proactive AI, ensuring that these measures adhere to ethical principles and legal requirements. Regulation should promote innovation while protecting public interests, providing a framework within which AI can develop safely and ethically.

In conclusion, the ethical implications of proactive measures in AI and robotics are complex and multifaceted. By balancing intervention and autonomy, ensuring transparency and fairness, preventing harm, engaging stakeholders, anticipating unintended consequences, respecting privacy, and implementing robust regulatory frameworks, we can navigate these ethical challenges effectively. Proactive measures, when designed and implemented ethically, hold the potential to enhance the positive impact of AI on society, ensuring that these technologies contribute to human well-being and the greater good.

Chapter 6: Understanding and Mitigating Threats

Risk Assessment

Identifying Potential Threats

In the rapidly evolving field of artificial intelligence and robotics, the task of identifying potential threats is paramount to ensuring the safety and integrity of AI systems. The process of threat identification requires a comprehensive understanding of the various ways in which AI can be compromised or misused, as well as the potential consequences of such threats. This understanding forms the basis for developing robust strategies to mitigate risks and safeguard human interests.

One of the primary challenges in identifying potential threats to AI systems lies in the sheer diversity and complexity of these threats. Threats can originate from a multitude of sources, including cyberattacks, data manipulation, and physical tampering. Cybersecurity threats, for instance, pose a significant risk as they can lead to unauthorized access, data breaches, and the disruption of AI operations. These attacks can be perpetrated by individuals, organizations, or even state actors with varying motives, from financial gain to political disruption.

Data manipulation is another critical threat to AI systems. AI relies heavily on large datasets for training and decision-making processes. If these datasets are corrupted, either through deliberate tampering or inadvertent errors, the resulting AI behavior can be unpredictable and potentially harmful. Ensuring the integrity of data, therefore, is a crucial aspect of threat identification. This involves not only securing data from external threats but also implementing rigorous validation and verification processes to detect and correct anomalies.

Physical tampering with AI hardware also presents a significant risk. As AI systems become more integrated into physical infrastructure, the potential for physical sabotage increases. This can include tampering with sensors, damaging critical components, or even introducing malicious hardware that can interfere with AI operations. Protecting AI systems from such physical threats requires robust security measures, regular maintenance, and monitoring to detect signs of tampering.

Beyond these tangible threats, AI systems also face ethical and societal challenges that can be considered threats to their successful implementation. Bias in AI algorithms, for example, can lead to unfair and discriminatory outcomes. This not only undermines the credibility and effectiveness of AI but also poses significant ethical dilemmas. Identifying and mitigating bias is thus a crucial component of threat identification. This involves scrutinizing the training data, the design of algorithms, and the deployment context to ensure fairness and equity.

Furthermore, the potential misuse of AI by malicious actors represents a significant threat. This can include the use of AI for surveillance, autonomous weapons, or misinformation campaigns. The dual-use nature of many AI technologies means that they can be repurposed for harmful activities. Identifying these misuse scenarios requires foresight and an understanding of the broader geopolitical and social landscape. It also necessitates collaboration with policymakers, ethicists, and other stakeholders to develop regulations and guidelines that prevent misuse while promoting beneficial applications.

The process of threat identification is not static; it requires continuous vigilance and adaptation. As AI technologies evolve, so do the threats. This dynamic nature of threats means that AI systems must be designed with flexibility and resilience in mind. Continuous monitoring, regular updates, and adaptive learning mechanisms are essential to ensure that AI systems can respond to new and emerging threats effectively.

In conclusion, identifying potential threats to AI systems is a multifaceted and ongoing process that is critical to the safe and ethical deployment of AI. By understanding the diverse nature of threats—from cybersecurity and data manipulation to physical tampering and ethical concerns—we can develop comprehensive strategies to mitigate risks. This proactive approach ensures that AI technologies can achieve their full potential in enhancing human capabilities while safeguarding against the myriad threats that could undermine their success.

Quantifying Risks

In the realm of artificial intelligence and robotics, quantifying risks is an essential aspect of ensuring the safe and ethical deployment of these technologies. Risk quantification involves the systematic assessment and measurement of potential threats and vulnerabilities associated with AI systems. This process is vital for developing strategies to mitigate risks and for making informed decisions about the deployment and management of AI technologies.

The first step in quantifying risks is to identify the various sources of potential harm. These can include technical failures, such as software bugs or hardware malfunctions, as well as security vulnerabilities that could be exploited by malicious actors. Additionally, the unintended consequences of AI decisions, such as biases in algorithmic outputs, pose significant risks that need to be quantified. Each of these sources of risk can have different probabilities and magnitudes of impact, which must be evaluated to understand the overall risk profile of an AI system.

Once potential sources of risk are identified, the next step is to assess their likelihood and potential impact. This assessment involves both qualitative and quantitative methods. Qualitative methods, such as expert judgment and scenario analysis, provide insights into the possible ways risks might manifest and their potential consequences. Quantitative methods, on the other hand, involve statistical analysis and modeling to estimate the probability of various risk events and their potential impacts. These methods can include techniques such as fault tree analysis, Bayesian networks, and Monte Carlo simulations, which help in creating a detailed and robust risk assessment.

A crucial aspect of quantifying risks is understanding the interdependencies and interactions between different risk factors. AI systems often operate within complex environments where multiple variables can influence outcomes. For example, a security breach might not only compromise data integrity but also affect system functionality and user trust. By modeling these interdependencies, we can gain a more comprehensive understanding of the overall risk landscape and identify critical points of vulnerability that need to be addressed.

In addition to assessing the probability and impact of risks, it is also important to consider the uncertainties associated with these assessments. Uncertainty can arise from various sources, including incomplete knowledge, variability in data, and the inherent unpredictability of complex systems. Quantifying uncertainty is essential for making informed decisions and for developing robust risk mitigation strategies. Techniques such as sensitivity analysis and uncertainty propagation can help in understanding how uncertainties affect the overall risk assessment and in identifying areas where more information or improved models are needed.

Effective risk quantification also involves continuous monitoring and updating of risk assessments. As AI systems evolve and new threats emerge, risk assessments need to be revisited and revised to reflect the current state of knowledge and technology. This dynamic approach to risk management ensures that AI systems remain resilient and capable of adapting to changing conditions. Continuous monitoring also allows for the early detection of potential issues, enabling proactive measures to mitigate risks before they escalate.

Moreover, the ethical implications of risk quantification cannot be overlooked. The process of assessing and quantifying risks must be conducted with transparency and accountability. Stakeholders, including AI developers, users, and regulatory bodies, need to be involved in the risk assessment process to ensure that all relevant perspectives are considered and that the resulting risk assessments are trusted and accepted. Ethical considerations also include the fair distribution of risks and benefits, ensuring that vulnerable populations are not disproportionately affected by the deployment of AI technologies.

In conclusion, quantifying risks in AI and robotics is a complex and multifaceted process that is essential for ensuring the safe and ethical deployment of these technologies. By systematically identifying, assessing, and managing risks, we can develop robust strategies to mitigate potential harms and maximize the benefits of AI systems. Continuous monitoring and ethical considerations further enhance the effectiveness of risk quantification, ensuring that AI technologies are deployed in a manner that is both responsible and beneficial to society.

Case Studies in Risk Assessment

In exploring the real-world application of risk assessment in artificial intelligence, case studies provide invaluable insights into how theoretical frameworks translate into practice. These studies illuminate the challenges and successes encountered in implementing robust risk management strategies, highlighting best practices and areas needing improvement.

One notable case study involves the deployment of autonomous vehicles (AVs). The introduction of AVs promised significant advancements in transportation efficiency and safety. However, the process of assessing and mitigating risks associated with these technologies has been complex. In one instance, a major technology firm conducted extensive testing of its autonomous vehicles in diverse environments to understand and address potential risks. The risk assessment focused on technical failures, such as sensor malfunctions and software errors, as well as external threats like cyberattacks and environmental factors. This comprehensive approach allowed the company to identify and prioritize high-risk scenarios, leading to the development of advanced safety protocols and fail-safe mechanisms. Despite these measures, an unfortunate incident occurred where an AV failed to detect a pedestrian, resulting in a fatal accident. This highlighted the importance of continuous monitoring and improvement of risk assessment processes, emphasizing that even well-prepared systems require ongoing evaluation and adaptation to address emerging threats.

Another illustrative case study is found in the healthcare sector, where AI has been integrated into diagnostic tools and patient management systems. One prominent example involved an AI system designed to assist radiologists in detecting anomalies in medical imaging. The initial risk assessment identified potential biases in the training data, which predominantly featured images from specific demographics, potentially leading to inaccurate diagnoses for underrepresented groups. To mitigate this risk, the developers expanded the dataset to include a more diverse range of images and implemented algorithms to adjust for demographic variations. Additionally, they conducted rigorous testing and validation phases to ensure the system's reliability and accuracy across different populations. This case study underscores the critical role of thorough data analysis and validation in managing risks associated with AI in healthcare, ensuring equitable and accurate outcomes for all patients.

The financial industry offers another pertinent example with the use of AI in fraud detection and risk management. A leading financial institution implemented an AI-driven system to monitor transactions and detect fraudulent activities. The initial risk assessment focused on the accuracy of the AI model, the potential for false positives and negatives, and the system's resilience to adversarial attacks. By employing advanced machine learning techniques and continuously updating the model with new data, the institution was able to enhance the system's detection capabilities. However, the assessment also revealed significant risks related to data privacy and regulatory compliance. To address these, the institution implemented robust data encryption methods and established strict access controls, ensuring that the system complied with all relevant regulations. This case study illustrates the necessity of balancing technical efficiency with regulatory and ethical considerations in AI risk management.

In the field of cybersecurity, AI has been employed to identify and respond to threats in real-time. A notable case involves a cybersecurity firm that developed an AI-based intrusion detection system (IDS). The risk assessment for this system included evaluating the effectiveness of the AI algorithms in detecting a wide range of cyber threats, the system's ability to operate under high-load conditions, and its resilience to evasion techniques employed by sophisticated attackers. The firm conducted extensive testing using simulated attack scenarios to refine the AI algorithms and enhance the IDS's performance. Additionally, the risk assessment process identified the need for human oversight to address complex and ambiguous threat scenarios that the AI might not handle effectively. By integrating human expertise with AI capabilities, the firm achieved a more comprehensive and adaptive approach to cybersecurity risk management.

These case studies demonstrate the multifaceted nature of risk assessment in AI applications. They highlight the importance of a thorough and continuous risk management process that includes identifying potential threats, assessing their likelihood and impact, and implementing effective mitigation strategies. The real-world experiences from these diverse sectors underscore the necessity of integrating technical, ethical, and regulatory considerations to ensure the safe and effective deployment of AI technologies. Through these examples, we can glean best practices and insights that inform the ongoing development and refinement of risk assessment frameworks in AI, ultimately contributing to more robust and trustworthy AI systems across various domains.

AI-driven Mitigation Strategies

Real-time Threat Detection

In the dynamic landscape of artificial intelligence, the ability to detect threats in real-time has become a critical component of maintaining system integrity and security. This capability is essential not only for safeguarding data and operations but also for ensuring the ethical deployment of AI technologies. Real-time threat detection involves the continuous monitoring and analysis of data to identify and respond to potential threats as they occur, thereby minimizing the risk of damage and enabling swift corrective actions.

The process of real-time threat detection in AI systems begins with the collection of vast amounts of data from various sources, including network traffic, user interactions, and system logs. This data is then analyzed using advanced algorithms and machine learning techniques to identify patterns and anomalies that could indicate a potential threat. Machine learning models are particularly effective in this context as they can be trained to recognize subtle and complex patterns that may not be apparent through traditional analysis methods. By continuously updating these models with new data, AI systems can adapt to evolving threat landscapes and improve their detection capabilities over time.

A key aspect of real-time threat detection is the ability to differentiate between normal and abnormal behavior. This involves establishing a baseline of normal system activity against which deviations can be measured. For instance, a sudden spike in network traffic or an unexpected change in user behavior could signal a potential security breach. AI systems use this baseline to flag anomalies for further investigation. This approach not only helps in identifying actual threats but also reduces the number of false positives, thereby improving the efficiency and effectiveness of the threat detection process.

In addition to anomaly detection, real-time threat detection systems employ various techniques such as signature-based detection, behavioral analysis, and predictive analytics. Signature-based detection involves comparing incoming data against known threat signatures, such as malware patterns or previously identified attack vectors. Behavioral analysis, on the other hand, focuses on understanding the typical behavior of users and systems to identify deviations that might indicate malicious activity. Predictive analytics uses historical data to forecast potential threats and proactively address them before they can cause harm.

The implementation of real-time threat detection also requires robust infrastructure and efficient data processing capabilities. High-speed data processing and real-time analytics are essential to ensure that threats are detected and addressed promptly. This necessitates the use of powerful computing resources and advanced data analytics platforms capable of handling large volumes of data and performing complex computations in real time. Additionally, the integration of AI systems with existing security frameworks and protocols is crucial for seamless threat detection and response.

Moreover, real-time threat detection is not limited to digital threats alone. In the context of autonomous systems and robotics, physical threats must also be considered. For example, an autonomous vehicle must be able to detect and respond to physical obstacles or hazards in real time to ensure the safety of passengers and pedestrians. This involves the use of sensors, cameras, and other monitoring devices to gather data about the vehicle's surroundings and the application of AI algorithms to process this data and make instant decisions.

Another critical element of real-time threat detection is the human-AI collaboration. While AI systems are adept at processing vast amounts of data and identifying potential threats, human oversight is essential for making complex ethical decisions and managing ambiguous situations. Security analysts and experts work alongside AI systems to validate alerts, investigate suspicious activities, and implement appropriate responses. This collaboration ensures that the strengths of both AI and human intelligence are leveraged to maintain a secure environment.

The importance of real-time threat detection extends beyond individual systems to the broader ecosystem in which AI operates. As AI technologies become more interconnected and integrated into various aspects of society, the potential impact of threats increases. Real-time threat detection helps to protect not only individual systems but also the overall network and infrastructure, thereby enhancing the resilience and reliability of AI technologies.

In conclusion, real-time threat detection is a vital component of modern AI systems, enabling them to operate securely and ethically in an increasingly complex threat landscape. By leveraging advanced algorithms, continuous monitoring, and human-AI collaboration, real-time threat detection systems can effectively identify and respond to potential threats, ensuring the integrity and safety of AI technologies. This proactive approach to security is essential for fostering trust in AI and enabling its safe and beneficial deployment across various domains.

Automated Response Systems

In the evolving landscape of artificial intelligence and robotics, the development and deployment of automated response systems are crucial for enhancing operational efficiency, security, and ethical compliance. These systems, designed to autonomously manage and respond to various scenarios, play a pivotal role in ensuring that AI and robotic technologies operate smoothly and safely in real-time environments.

Automated response systems are engineered to detect anomalies, threats, and other significant events, triggering appropriate actions without human intervention. This capability is particularly vital in high-stakes domains such as cybersecurity, healthcare, and autonomous transportation, where swift and accurate responses can prevent potential disasters and mitigate risks effectively. These systems rely on advanced algorithms and machine learning models that continuously analyze incoming data, identify patterns, and predict possible issues before they escalate.

In cybersecurity, automated response systems are indispensable for protecting networks and data from malicious activities. These systems monitor network traffic, identify suspicious activities, and automatically implement countermeasures to neutralize threats. For instance, when an intrusion detection system flags an unusual access attempt, the automated response system can immediately isolate the compromised network segment, block the attacker's IP address, and alert security personnel. This rapid response minimizes the potential damage and helps maintain the integrity and confidentiality of critical data.

In the healthcare sector, automated response systems enhance patient care and safety by continuously monitoring vital signs and other health indicators. These systems can detect early signs of medical emergencies, such as cardiac arrest or respiratory failure, and automatically alert healthcare providers while initiating predefined protocols. For example, if a patient's heart rate drops to a critical level, the system can summon emergency medical staff, activate life-support systems, and provide real-time data to assist in immediate medical intervention. Such automation not only improves response times but also reduces the burden on healthcare professionals, allowing them to focus on more complex tasks.

The deployment of automated response systems in autonomous vehicles exemplifies their critical role in ensuring safety and reliability. These systems constantly analyze data from sensors and cameras to detect obstacles, traffic signals, and other vehicles, making real-time decisions to navigate safely. In emergency situations, such as the sudden appearance of an obstacle, the automated system can swiftly execute maneuvers to avoid collisions, thereby protecting passengers and pedestrians. The reliability of these systems is enhanced by their ability to learn from vast amounts of driving data, improving their decision-making capabilities over time.

Despite their numerous benefits, the implementation of automated response systems also raises significant ethical and technical challenges. One major concern is the potential for false positives and negatives, where the system may either overreact to benign events or fail to respond to genuine threats. Addressing this issue requires rigorous testing and validation of the underlying algorithms to ensure their accuracy and reliability. Additionally, transparency in how these systems make decisions is crucial for gaining public trust and ensuring accountability.

Another critical challenge is balancing the autonomy of these systems with the need for human oversight. While automated systems can handle many tasks independently, human intervention remains necessary for complex and ambiguous situations that require ethical judgment and contextual understanding. Therefore, a hybrid approach, where automated systems and human operators work collaboratively, is often the most effective strategy. This approach leverages the strengths of both automation and human intuition, providing a robust framework for managing emergencies and other critical scenarios.

The integration of automated response systems also demands robust infrastructure and regulatory frameworks to ensure their safe and ethical deployment. This includes developing standards for data privacy and security, establishing protocols for system maintenance and updates, and creating guidelines for the ethical use of AI and robotics. Policymakers, industry leaders, and researchers must collaborate to address these challenges, fostering an environment where automated response systems can thrive while safeguarding public interests.

In conclusion, automated response systems are a cornerstone of modern AI and robotics, offering significant advantages in terms of efficiency, safety, and ethical compliance. By enabling real-time detection and response to various events, these systems enhance the capabilities of AI technologies and ensure their reliable and ethical operation. As these systems continue to evolve, addressing the associated challenges through rigorous testing, transparent decision-making, and effective human-AI collaboration will be essential for realizing their full potential and ensuring their beneficial impact on society.

Human-AI Collaboration

In the advancing realm of artificial intelligence and robotics, the synergy between human intelligence and AI systems is becoming increasingly critical. This collaborative approach combines the strengths of human creativity, intuition, and ethical judgment with the computational power and data-processing capabilities of AI. Human-AI collaboration aims to enhance decision-making processes, improve efficiency, and ensure that AI technologies are applied ethically and responsibly across various domains.

At the heart of human-AI collaboration is the recognition that while AI systems excel at handling large volumes of data and performing repetitive tasks, they lack the nuanced understanding and ethical considerations that humans bring to complex situations. Therefore, a collaborative framework allows humans to guide AI systems, ensuring that their actions align with societal values and ethical standards. This partnership is particularly important in areas such as healthcare, where AI can assist in diagnosing diseases and suggesting treatments, but human doctors must interpret these recommendations within the context of individual patient needs and ethical considerations.

In the medical field, human-AI collaboration can lead to significant advancements in patient care. AI systems can analyze medical images, identify patterns that might be missed by human eyes, and provide preliminary diagnoses. However, these AI-generated insights are most effective when reviewed by medical professionals who can consider additional factors such as patient history, lifestyle, and potential treatment side effects. For example, an AI system might detect early signs of cancer in imaging scans, but a human oncologist is needed to develop a comprehensive treatment plan tailored to the patient's specific circumstances. This collaborative approach not only improves diagnostic accuracy but also ensures that medical decisions are made with a holistic understanding of the patient's condition.

In the realm of autonomous vehicles, human-AI collaboration is essential for ensuring safety and reliability. While AI systems can navigate complex driving environments and respond to real-time traffic conditions, human oversight remains crucial for managing unexpected situations and making ethical decisions. For instance, in scenarios where an autonomous vehicle must choose between two potentially harmful outcomes, human input is necessary to guide the AI system in making ethically sound decisions. This collaborative approach can be seen in semi-autonomous driving systems, where drivers are required to stay alert and take control when necessary, ensuring a balance between automation and human judgment.

Human-AI collaboration also plays a pivotal role in cybersecurity, where AI systems are used to detect and respond to threats in real-time. AI can monitor network traffic, identify anomalies, and predict potential security breaches based on patterns in the data. However, cybersecurity experts are needed to interpret these findings, assess the severity of threats, and decide on the appropriate responses. This collaboration ensures that the AI system's recommendations are grounded in a broader understanding of the organization's security posture and strategic objectives. For example, while an AI system might flag a particular network activity as suspicious, a human expert can determine whether it is a false positive or a genuine threat, thereby refining the system's accuracy over time.

In the financial industry, human-AI collaboration enhances risk management and decision-making processes. AI systems can analyze market trends, predict economic shifts, and identify potential investment opportunities. Financial analysts then use these insights to make informed decisions, considering factors that the AI might overlook, such as geopolitical developments or regulatory changes. This collaborative approach allows for more comprehensive risk assessments and strategic planning, combining the strengths of AI in data analysis with human expertise in interpreting and applying these insights.

The ethical dimension of human-AI collaboration cannot be overstated. As AI systems become more integrated into decision-making processes, it is crucial to ensure that they operate within ethical guidelines. Humans must oversee AI applications to prevent biases, ensure fairness, and protect individual rights. This oversight includes developing and enforcing ethical standards for AI development and deployment, conducting regular audits to identify and mitigate biases, and engaging diverse stakeholders in the AI governance process. For example, in hiring processes where AI is used to screen candidates, human oversight is necessary to ensure that the AI system does not perpetuate existing biases and that hiring decisions are made fairly and transparently.

In conclusion, human-AI collaboration is essential for harnessing the full potential of AI technologies while ensuring their ethical and responsible use. By combining human judgment, creativity, and ethical considerations with the computational power of AI, we can achieve better outcomes across various domains, from healthcare and autonomous driving to cybersecurity and finance. This collaborative approach not only enhances the effectiveness of AI systems but also ensures that their deployment aligns with societal values and ethical principles, fostering trust and confidence in AI technologies.

Continuous Monitoring

Importance of Monitoring

The continuous monitoring of AI and robotic systems is essential for maintaining their reliability, safety, and ethical alignment. As these technologies become more integrated into critical aspects of daily life, the need to ensure their consistent performance and adherence to ethical standards becomes paramount. Monitoring serves multiple purposes, from detecting anomalies and preventing malfunctions to ensuring compliance with regulatory requirements and ethical guidelines.

One of the primary reasons for continuous monitoring is to detect and address anomalies in real-time. AI systems and robots operate in dynamic environments where unexpected changes can occur. These anomalies could be due to software bugs, hardware failures, or external factors such as cyberattacks. By constantly monitoring system performance and environmental inputs, anomalies can be detected early, allowing for prompt corrective actions. This early detection is crucial in preventing minor issues from escalating into major failures, which could have significant consequences, especially in critical applications like healthcare, transportation, and security.

In healthcare, for instance, monitoring AI systems that assist in patient diagnosis and treatment is vital. These systems analyze vast amounts of medical data to provide recommendations. Any deviation from expected performance could lead to incorrect diagnoses or inappropriate treatment plans. Continuous monitoring ensures that any discrepancies are identified and corrected swiftly, maintaining the accuracy and reliability of medical AI systems. This is particularly important when dealing with life-threatening conditions where timely and accurate intervention is crucial.

In the realm of autonomous vehicles, monitoring systems are critical for ensuring passenger safety. Autonomous vehicles rely on complex algorithms and a multitude of sensors to navigate and make decisions. Constant monitoring of these systems helps in identifying sensor failures, software glitches, or unexpected obstacles in real time. For example, if a sensor fails to detect a pedestrian, the monitoring system can trigger an emergency response, such as slowing down or stopping the vehicle. This continuous oversight ensures that autonomous vehicles operate safely and reliably, mitigating the risks associated with autonomous driving.

Cybersecurity also benefits significantly from continuous monitoring. AI systems are increasingly targeted by cyberattacks due to their critical roles in various sectors. Monitoring AI systems for signs of cyber threats, such as unusual network traffic or unauthorized access attempts, helps in identifying and neutralizing threats before they can cause harm. This proactive approach to cybersecurity is essential for protecting sensitive data and maintaining the integrity of AI systems. Moreover, continuous monitoring can help in detecting and mitigating insider threats, where authorized users might misuse their access privileges.

Ethical compliance is another critical aspect of monitoring AI systems. As AI technologies are deployed in various fields, ensuring that they operate within ethical boundaries is essential. Continuous monitoring helps in identifying instances where AI systems might deviate from ethical guidelines, such as exhibiting biased behavior or making decisions that could harm individuals or groups. For example, in hiring processes, AI systems must be monitored to ensure they do not discriminate against candidates based on race, gender, or other protected characteristics. By maintaining constant oversight, organizations can ensure that their AI systems uphold ethical standards and promote fairness and equality.

Monitoring also plays a vital role in compliance with regulatory requirements. Various industries have specific regulations that govern the use of AI and robotics. Continuous monitoring ensures that these systems comply with relevant laws and standards, avoiding legal issues and potential penalties. For instance, in the financial sector, AI systems used for trading must comply with regulations related to market manipulation and insider trading. Monitoring these systems ensures that they operate within legal boundaries and maintain market integrity.

Furthermore, continuous monitoring supports the iterative improvement of AI systems. By analyzing performance data over time, developers can identify areas for improvement and optimize algorithms and processes. This iterative approach leads to more robust and efficient AI systems that can better serve their intended purposes. For example, in industrial automation, monitoring production lines helps in identifying bottlenecks and inefficiencies, allowing for continuous improvement of manufacturing processes.

In conclusion, the importance of monitoring AI and robotic systems cannot be overstated. It ensures the reliability, safety, and ethical compliance of these technologies, enabling them to operate effectively in various critical applications. Through continuous monitoring, anomalies can be detected and addressed promptly, cybersecurity threats can be mitigated, and compliance with ethical and regulatory standards can be maintained. As AI and robotics continue to evolve and integrate into more aspects of life, robust monitoring practices will be essential for harnessing their full potential while safeguarding human interests.

Tools and Technologies

As artificial intelligence and robotics continue to advance, the development and implementation of various tools and technologies are crucial for ensuring their safe, efficient, and ethical operation. These tools encompass a wide range of applications, from enhancing machine learning algorithms to ensuring robust cybersecurity measures and improving human-AI interaction. Understanding and leveraging these technologies is essential for maximizing the potential of AI while mitigating associated risks.

One of the foundational technologies in AI is machine learning, particularly deep learning. Deep learning algorithms, modeled after the human brain's neural networks, allow AI systems to process large amounts of data and identify patterns with high accuracy. These algorithms are used in various applications, such as image and speech recognition, natural language processing, and predictive analytics. For instance, in healthcare, deep learning algorithms can analyze medical images to detect early signs of diseases like cancer, providing valuable support to medical professionals. However, developing these algorithms requires vast datasets and significant computational power, necessitating the use of advanced hardware like Graphics Processing Units (GPUs) and specialized software frameworks such as TensorFlow and PyTorch.

In addition to machine learning, natural language processing (NLP) is a critical area of AI that enables machines to understand and interact with human language. NLP technologies are used in virtual assistants, chatbots, and automated translation services. These tools rely on sophisticated algorithms to process and interpret human language, enabling more natural and intuitive interactions between humans and machines. For example, virtual assistants like Siri and Alexa use NLP to understand user commands and provide relevant responses, enhancing user experience and accessibility.

Cybersecurity is another vital area where advanced tools and technologies are essential. As AI systems become more integrated into critical infrastructure, the need to protect them from cyber threats becomes paramount. Advanced cybersecurity tools employ AI and machine learning to detect and respond to threats in real-time. These tools can identify patterns indicative of cyberattacks, such as unusual network traffic or unauthorized access attempts, and initiate automated responses to mitigate these threats. For instance, AI-driven intrusion detection systems can monitor network activity continuously and alert security teams to potential breaches, enabling swift action to protect sensitive data and maintain system integrity.

In the context of autonomous systems, such as self-driving cars, sensor technology and computer vision play a crucial role. These systems rely on a combination of cameras, LiDAR, radar, and ultrasonic sensors to perceive their environment and make real-time decisions. Advanced algorithms process the data from these sensors to identify objects, predict their movements, and navigate safely. The integration of these technologies ensures that autonomous vehicles can operate reliably and safely in complex and dynamic environments. Moreover, continuous advancements in sensor technology and computational power are enhancing the capabilities and safety of autonomous systems, bringing us closer to widespread adoption.

Human-AI interaction is another area where advanced tools are making significant strides. Developing interfaces that facilitate seamless and intuitive interaction between humans and AI systems is critical for maximizing the potential of these technologies. Tools such as augmented reality (AR) and virtual reality (VR) provide immersive experiences that can enhance human-AI collaboration. For example, in industrial settings, AR can be used to overlay digital information onto the physical world, assisting workers in complex assembly tasks by providing real-time guidance and feedback. Similarly, VR can be used for training and simulation purposes, allowing individuals to practice and hone their skills in a safe and controlled environment.

Ethical considerations are integral to the development and deployment of AI tools and technologies. Ensuring that AI systems operate within ethical boundaries requires the implementation of frameworks and guidelines that promote fairness, transparency, and accountability. Tools for bias detection and mitigation are essential for ensuring that AI systems do not perpetuate existing biases or create new forms of discrimination. For instance, algorithms can be developed to analyze training data and identify potential biases, allowing developers to address these issues proactively. Additionally, transparency tools, such as explainable AI (XAI), are being developed to provide insights into how AI systems make decisions, fostering trust and understanding among users.

In conclusion, the advancement of AI and robotics is heavily dependent on the continuous development and integration of sophisticated tools and technologies. From machine learning and NLP to cybersecurity and human-AI interaction, these tools are essential for ensuring the safe, efficient, and ethical operation of AI systems. As we continue to innovate and refine these technologies, it is crucial to address the associated challenges and ethical considerations, ensuring that AI serves humanity in the most beneficial and responsible way possible. By leveraging these tools effectively, we can unlock the full potential of AI and robotics, transforming various aspects of our lives and creating a future that is both technologically advanced and ethically sound.

Case Studies in Continuous Monitoring

In the field of artificial intelligence and robotics, continuous monitoring is critical for maintaining system integrity, safety, and performance. Various case studies highlight the importance of implementing robust monitoring practices and demonstrate the significant impact these practices can have across different sectors. These real-world examples illustrate how continuous monitoring can prevent failures, improve efficiency, and ensure compliance with ethical and regulatory standards.

One prominent case study involves the use of continuous monitoring in the healthcare sector, particularly in the management of intensive care units (ICUs). Advanced AI systems are employed to continuously monitor patients' vital signs and other health metrics. These systems utilize algorithms to detect early signs of deterioration, allowing healthcare providers to intervene before a patient's condition worsens. For example, in a leading hospital, an AI-driven monitoring system successfully reduced the incidence of sepsis by identifying early warning signs that were previously undetectable through traditional methods. This early detection allowed medical staff to administer timely treatments, significantly improving patient outcomes and reducing mortality rates. The continuous monitoring system not only enhanced patient care but also demonstrated the potential of AI to transform healthcare practices through proactive and precise interventions.

In the realm of autonomous vehicles, continuous monitoring is essential for ensuring safety and reliability. A well-documented case study involves a fleet of autonomous delivery robots deployed in urban environments. These robots are equipped with a variety of sensors, including cameras, LiDAR, and ultrasonic sensors, which continuously monitor their surroundings to navigate safely. The monitoring system tracks the robots' movements, detects obstacles, and ensures compliance with traffic rules. In one instance, continuous monitoring allowed the system to identify a malfunction in one of the sensors, which could have led to a collision. The monitoring system promptly flagged the issue, enabling the robot to be taken offline for maintenance before any harm occurred. This example underscores the importance of continuous monitoring in identifying and mitigating risks in real time, thereby ensuring the safe operation of autonomous systems.

Cybersecurity is another domain where continuous monitoring proves indispensable. In a major financial institution, AI-powered monitoring tools are used to oversee network activity and detect potential cyber threats. These tools analyze vast amounts of data to identify unusual patterns that might indicate a security breach. In one notable incident, the continuous monitoring system detected a series of unauthorized access attempts targeting the institution's critical infrastructure. The system flagged the suspicious activity and triggered automated defense mechanisms, including isolating the affected network segment and notifying the security team. The timely response prevented a potentially devastating data breach, protecting sensitive financial information and maintaining the integrity of the institution's operations. This case study highlights how continuous monitoring is vital for early threat detection and rapid response in protecting against cyberattacks.

In manufacturing, continuous monitoring is employed to enhance productivity and prevent equipment failures. A leading automotive manufacturer implemented an AI-based monitoring system to oversee its assembly lines. The system continuously collects data from various sensors installed on the machinery, analyzing it to detect signs of wear and tear or potential malfunctions. In one scenario, the monitoring system identified an abnormal vibration pattern in a critical component of the assembly line. By detecting this anomaly early, the system prompted maintenance staff to inspect and repair the component before it failed completely. This proactive approach minimized downtime, avoided costly repairs, and ensured that the production schedule remained on track. The case study demonstrates how continuous monitoring can improve operational efficiency and reduce the risk of unexpected equipment failures in industrial settings.

In the context of environmental monitoring, AI systems are used to continuously track air and water quality, providing valuable data for managing public health and safety. In a coastal city, an AI-powered monitoring network was deployed to monitor water quality in real time. The system collected data from various sensors placed in the water, analyzing it to detect pollutants and harmful substances. When the monitoring system detected a rise in contamination levels, it immediately alerted the city's environmental protection agency, which took swift action to identify and mitigate the pollution source. This continuous monitoring not only helped protect the local ecosystem but also ensured the safety of the city's residents by preventing exposure to hazardous water conditions.

These case studies across different sectors illustrate the profound impact of continuous monitoring on enhancing safety, efficiency, and reliability. Whether in healthcare, autonomous vehicles, cybersecurity, manufacturing, or environmental protection, continuous monitoring systems provide critical insights that enable proactive management and timely interventions. By leveraging advanced AI technologies for continuous monitoring, organizations can anticipate and address potential issues before they escalate, ensuring optimal performance and safeguarding human well-being. The integration of continuous monitoring into AI and robotic systems represents a significant advancement in technology, fostering a safer and more efficient future across various domains.

Chapter 7: Strategies for Deterrence, Containment, and Elimination

Deterrence Strategies

AI in Deterrence

Artificial intelligence has emerged as a significant factor in modern deterrence strategies, reshaping how nations approach security and conflict prevention. The integration of AI into military and defense systems enhances the ability to predict, detect, and respond to potential threats more swiftly and accurately than ever before. AI systems can analyze vast amounts of data to identify patterns and anomalies that may indicate hostile activities, providing critical insights that inform decision-making processes. This advanced capability allows for the development of more effective deterrence measures, reducing the likelihood of conflicts escalating into full-scale wars.

The use of AI in deterrence is not limited to mere data analysis. Autonomous systems, such as drones and unmanned vehicles, equipped with AI, can conduct surveillance, reconnaissance, and even offensive operations with minimal human intervention. These systems offer significant strategic advantages, including the ability to operate in environments that are too dangerous for human personnel and to respond to threats in real-time. The speed and precision of AI-driven systems ensure that potential adversaries are aware of the high risks and costs associated with aggressive actions, thereby reinforcing the deterrence posture of a nation.

Moreover, AI can play a crucial role in cybersecurity, a critical component of national defense in the digital age. AI-powered tools can detect and counteract cyber threats with unprecedented speed, protecting critical infrastructure and sensitive information from malicious attacks. By safeguarding these assets, AI strengthens a nation's overall security framework, further deterring adversaries from attempting cyber incursions.

However, the integration of AI into deterrence strategies also raises several ethical and strategic concerns. The potential for autonomous weapons to make life-and-death decisions without human oversight is a significant ethical dilemma. Ensuring that AI systems are used responsibly and within the bounds of international law is crucial to maintaining global stability. Additionally, the reliance on AI introduces new vulnerabilities, such as the risk of AI systems being hacked or manipulated by adversaries, which could lead to unintended escalations or conflicts.

In conclusion, the application of artificial intelligence in deterrence strategies represents a transformative shift in how nations protect their interests and maintain global stability. By enhancing the ability to predict, detect, and respond to threats, AI contributes to more effective deterrence and a reduction in the likelihood of conflicts. However, the ethical and strategic challenges associated with AI integration must be carefully managed to ensure that these technologies are used to promote peace and security responsibly.

Psychological Deterrence

The concept of psychological deterrence revolves around influencing the perceptions and decision-making processes of potential adversaries to prevent hostile actions. This approach relies on understanding the cognitive and emotional factors that drive behavior. By shaping the environment in which decisions are made, psychological deterrence aims to instill doubt, fear, or a sense of futility in the minds of adversaries, thereby discouraging them from pursuing aggressive actions. Effective psychological deterrence involves a deep understanding of the adversary's values, beliefs, and decision-making frameworks.

One of the key elements of psychological deterrence is the strategic communication of threats and consequences. By clearly conveying the costs and risks associated with hostile actions, a state can create a mental calculus for the adversary where the perceived negatives of acting aggressively outweigh any potential benefits. This involves not only the explicit articulation of military and economic repercussions but also more subtle cues and signals that convey resolve and capability.

Another critical aspect is the demonstration of capability and readiness. Through military exercises, strategic deployments, and technological displays, a state can project strength and preparedness. These actions serve to reinforce the communicated threats, making them more credible in the eyes of potential adversaries. The goal is to create a perception of invincibility or overwhelming superiority that deters adversaries from challenging the status quo.

Psychological deterrence also benefits from a nuanced understanding of cultural and psychological factors that influence decision-making. Different cultures and political systems may respond differently to threats and incentives. Tailoring deterrence strategies to the specific psychological and cultural context of the adversary increases their effectiveness. For example, some regimes might be more influenced by demonstrations of international solidarity and legitimacy, while others might be more concerned with domestic perceptions of strength and stability.

In addition to state actors, psychological deterrence can be applied to non-state actors and terrorist organizations. Here, the focus may shift to undermining the ideological and motivational foundations that support hostile actions. By disrupting the narratives and beliefs that justify violence, psychological deterrence can weaken the resolve of these groups and reduce their operational effectiveness.

The integration of advanced technologies such as artificial intelligence and big data analytics has further enhanced the capability to implement psychological deterrence. These technologies allow for more precise and tailored messaging, as well as real-time monitoring of adversary responses. This dynamic feedback loop enables continuous adjustment of deterrence strategies to maintain their effectiveness.

In conclusion, psychological deterrence is a multifaceted approach that leverages an understanding of human cognition and behavior to prevent hostile actions. By strategically communicating threats, demonstrating capability, and tailoring strategies to specific cultural contexts, states can effectively dissuade adversaries from pursuing aggressive policies. The integration of advanced technologies further amplifies the potential of psychological deterrence, making it a critical component of modern security strategies.

Case Studies in Effective Deterrence

Effective deterrence is often best understood through concrete examples where strategies have been successfully applied to prevent conflicts and maintain stability. One notable case is the Cuban Missile Crisis of 1962, which stands as a prime example of deterrence in action during the Cold War. The United States discovered Soviet missiles in Cuba, prompting a tense standoff that brought the world to the brink of nuclear war. Through a combination of naval blockade, diplomatic negotiations, and the clear communication of potential military responses, the U.S. successfully pressured the Soviet Union to remove the missiles, thereby averting a catastrophic conflict. This incident highlights how credible threats and measured responses can compel adversaries to reconsider aggressive actions.

Another illustrative case is the ongoing deterrence dynamics between North and South Korea. South Korea, backed by the United States, employs a robust combination of military readiness, economic sanctions, and diplomatic efforts to deter North Korean aggression. Despite periodic escalations, this multifaceted approach has largely succeeded in preventing full-scale war on the Korean Peninsula. The situation underscores the importance of a comprehensive deterrence strategy that includes military, economic, and diplomatic tools to address complex security challenges.

In more recent times, the cyber domain has become a critical arena for deterrence. The 2016 U.S. presidential election interference by Russia highlighted the vulnerabilities in cyber infrastructure and the need for effective deterrence strategies in this sphere. Since then, the U.S. has developed a range of cyber deterrence measures, including the threat of retaliatory cyberattacks, public attribution of malicious activities, and international cooperation to strengthen cybersecurity norms. These efforts aim to dissuade adversaries from conducting cyber operations that threaten national security and democratic processes.

The deterrence of terrorism also presents unique challenges and requires tailored strategies. The U.S. and its allies have employed a combination of intelligence operations, targeted military strikes, and efforts to undermine terrorist ideologies to deter groups like ISIS and Al-Qaeda. The success of these efforts often hinges on the ability to disrupt terrorist networks, diminish their operational capabilities, and counter the narratives that support their activities. By weakening the appeal and effectiveness of terrorist organizations, states can reduce the likelihood of attacks and enhance overall security.

Furthermore, the use of economic sanctions as a tool of deterrence is evident in the case of Iran's nuclear program. The international community, led by the United States, imposed stringent economic sanctions on Iran to compel it to negotiate limits on its nuclear activities. The resulting Joint Comprehensive Plan of Action (JCPOA) demonstrated how economic pressure, combined with diplomatic engagement, can achieve significant security objectives without resorting to military force.

These case studies illustrate the diverse applications of deterrence across different domains and conflicts. They emphasize the need for credible threats, the importance of clear communication, and the role of comprehensive strategies that integrate military, economic, and diplomatic elements. Effective deterrence not only prevents immediate conflicts but also contributes to long-term stability by shaping the strategic calculations of potential adversaries. As the nature of threats continues to evolve, so too must the approaches to deterrence, incorporating new technologies and adapting to changing geopolitical landscapes to maintain peace and security.

Containment Techniques

Technological Containment

In the contemporary landscape of international security, technological containment has become a critical strategy for managing and mitigating threats. This approach involves leveraging advanced technologies to monitor, control, and neutralize potential dangers before they can escalate into significant crises. One key area of technological containment is the development and deployment of sophisticated surveillance systems. These systems, often enhanced by artificial intelligence and machine learning algorithms, can process vast amounts of data in real-time, identifying patterns and anomalies that may indicate emerging threats. For example, satellite imagery, drone reconnaissance, and cyber intelligence provide comprehensive situational awareness, allowing for timely and informed decision-making.

Additionally, the integration of cyber defense mechanisms into national security strategies exemplifies the role of technology in containment. Cyber threats pose unique challenges due to their covert nature and the speed at which they can be executed. Advanced cyber defense systems utilize artificial intelligence to detect and respond to cyber intrusions instantaneously, mitigating potential damage and preventing the spread of malicious activities. These systems are designed to protect critical infrastructure, such as power grids, communication networks, and financial systems, from both state-sponsored cyber-attacks and non-state actors.

Furthermore, technological containment extends to the realm of missile defense. Modern missile defense systems incorporate radar, satellites, and ground-based interceptors to detect, track, and neutralize incoming missiles. These systems, exemplified by technologies like the Aegis Ballistic Missile Defense and the Terminal High Altitude Area Defense (THAAD), provide a shield against potential missile attacks, enhancing a nation's defensive posture and deterring adversaries from contemplating missile strikes.

The application of containment strategies is also evident in the use of autonomous systems for border security. Unmanned aerial vehicles (UAVs), ground sensors, and automated surveillance towers are deployed to monitor and secure borders, detecting and preventing unauthorized crossings and smuggling activities. These technologies offer persistent surveillance capabilities, covering vast and remote areas that would be challenging to monitor with human patrols alone.

Moreover, the concept of technological containment is not limited to military applications. In public health, technologies such as contact tracing apps and genomic sequencing are used to contain the spread of infectious diseases. These tools enable health authorities to track outbreaks, identify transmission chains, and implement targeted interventions to prevent widespread contagion. The COVID-19 pandemic has underscored the importance of such technologies in managing global health crises.

In conclusion, technological containment represents a multifaceted approach to addressing contemporary security challenges. By harnessing advanced technologies in surveillance, cyber defense, missile interception, border security, and public health, nations can effectively monitor and neutralize potential threats. This proactive and preventive strategy not only enhances national security but also contributes to global stability by reducing the risk of conflicts and crises. As technology continues to evolve, so too will the methods and tools of containment, ensuring that nations remain equipped to handle the dynamic and complex security landscape of the future.

Physical Containment

The concept of physical containment in international security involves the use of tangible measures to restrict the spread of potentially harmful elements, such as nuclear materials, biological agents, and hazardous substances. This strategy is critical in preventing the proliferation of weapons of mass destruction and ensuring global stability. One prominent example is the implementation of nuclear non-proliferation protocols. These include strict regulatory frameworks and international agreements like the Treaty on the Non-Proliferation of Nuclear Weapons (NPT), which aims to prevent the spread of nuclear weapons and promote peaceful uses of nuclear energy.

Physical containment also encompasses the secure storage and transport of nuclear materials. Facilities such as secure vaults and containment buildings are designed to withstand natural disasters and potential attacks, ensuring that nuclear materials remain inaccessible to unauthorized individuals or groups. Moreover, sophisticated monitoring systems, including sensors and surveillance technologies, are employed to detect and respond to any breaches in security.

In the realm of biological containment, facilities known as biosafety laboratories are utilized to safely handle and research dangerous pathogens. These laboratories are classified into different biosafety levels, each with specific containment measures appropriate for the type of pathogens being studied. The highest level, Biosafety Level 4 (BSL-4), includes airtight suits, specialized air filtration systems, and rigorous decontamination procedures to prevent any accidental release of infectious agents.

The containment of hazardous substances extends to chemical weapons as well. International treaties like the Chemical Weapons Convention (CWC) mandate the destruction of chemical weapon stockpiles and the prohibition of their use, development, and transfer. Compliance is monitored through inspections and verification measures, ensuring that member states adhere to their obligations.

Physical containment measures are also crucial in the context of environmental protection. The management of toxic waste, for example, involves the use of secure landfills and containment facilities to prevent contamination of soil and water resources. Advanced technologies and engineering solutions are employed to monitor and mitigate the impact of hazardous waste on the environment.

In the field of public health, physical containment plays a vital role in controlling the spread of infectious diseases. Quarantine facilities, isolation wards, and vaccination campaigns are some of the measures used to contain outbreaks and protect populations from widespread contagion. The COVID-19 pandemic has highlighted the importance of effective physical containment strategies, including the establishment of quarantine zones and the rapid deployment of medical resources.

Overall, physical containment is a multifaceted approach that integrates regulatory frameworks, technological solutions, and international cooperation to manage and mitigate risks associated with dangerous materials and pathogens. By implementing robust containment measures, the international community can effectively prevent the proliferation of weapons of mass destruction, protect the environment, and safeguard public health, thereby contributing to global security and stability.

Case Studies in Containment

Examining various case studies in containment reveals the diverse strategies employed to manage and neutralize potential threats effectively. The Chernobyl disaster in 1986 is a stark example of physical containment in response to a catastrophic event. Following the explosion at the nuclear power plant, an exclusion zone was established, and a massive concrete sarcophagus was constructed to encase the reactor and contain the spread of radioactive materials. This effort was later supplemented by the New Safe Confinement structure, completed in 2016, which further encapsulated the reactor, preventing the release of radiation and enabling the safe dismantling of the original sarcophagus. The Chernobyl containment measures illustrate the critical importance of quick, decisive action and the use of advanced engineering solutions to mitigate the effects of nuclear disasters.

Another significant case is the Ebola outbreak in West Africa between 2014 and 2016. The containment strategy for this epidemic involved a combination of quarantine measures, international medical assistance, and the establishment of treatment centers equipped with stringent infection control protocols. Health organizations such as the World Health Organization (WHO) and Médecins Sans Frontières (MSF) played pivotal roles in coordinating efforts to isolate and treat infected individuals, trace contacts, and educate the public on prevention methods. The eventual development and deployment of an effective Ebola vaccine also contributed to containing the outbreak and preventing its spread to other regions. This case underscores the importance of global cooperation, rapid response, and the integration of medical and logistical strategies in managing public health crises.

The Fukushima Daiichi nuclear disaster in 2011 provides another example of containment efforts in the wake of a major nuclear incident. After a tsunami caused by a massive earthquake led to reactor meltdowns, immediate efforts were made to stabilize the reactors and contain radioactive releases. This included the injection of seawater to cool the reactors, the construction of a temporary containment barrier, and the long-term plan to decommission the reactors and manage radioactive waste. The ongoing decontamination and decommissioning efforts highlight the complexities and extended timeframes often required in the containment of nuclear accidents, emphasizing the need for sustained commitment and technological innovation.

Additionally, the implementation of the Strategic Arms Reduction Treaty (START) between the United States and the Soviet Union (later Russia) serves as a key example of containment through international diplomacy and arms control. The treaty, first signed in 1991 and followed by subsequent agreements, aimed to reduce the number of nuclear weapons and delivery systems, thus containing the potential for nuclear conflict. Verification mechanisms, including on-site inspections and data exchanges, ensured compliance and built trust between the two superpowers. This case demonstrates how diplomatic negotiations and legal frameworks can effectively contribute to global security by containing the proliferation and deployment of nuclear arsenals.

In summary, these case studies highlight the varied and multifaceted approaches to containment across different domains, including nuclear safety, public health, and international security. The successful containment of threats often requires a combination of immediate tactical responses, long-term strategic planning, international cooperation, and technological innovation. By learning from these examples, future efforts in containment can be better prepared to address and mitigate emerging risks, ensuring a safer and more stable global environment.

Elimination Approaches

Ethical Considerations

Ethical considerations in the deployment and use of advanced technologies are paramount, especially in fields that significantly impact society, such as artificial intelligence, biotechnology, and cybersecurity. As technological capabilities expand, so do the potential risks and moral dilemmas associated with their use. The ethical landscape is shaped by the need to balance innovation with responsibility, ensuring that technological advancements contribute positively to humanity while minimizing harm.

One of the primary ethical concerns is the potential for bias in AI systems. Algorithms trained on biased data can perpetuate and even amplify existing inequalities, leading to unfair outcomes in areas such as hiring, law enforcement, and lending. Ensuring that AI systems are transparent, accountable, and developed with diverse datasets is crucial in mitigating these risks. Moreover, there is a pressing need for ethical guidelines and regulatory frameworks to govern the use of AI, ensuring that these technologies are used in ways that respect human rights and promote social justice.

In biotechnology, the manipulation of genetic material raises significant ethical questions. Advances in gene editing, particularly CRISPR technology, hold the promise of curing genetic diseases but also pose risks of unintended consequences and ethical dilemmas regarding genetic enhancement. The potential for creating "designer babies" and the long-term impacts on genetic diversity and societal inequality must be carefully considered. International cooperation and stringent ethical standards are essential in guiding the responsible use of these powerful technologies.

Cybersecurity also presents a complex ethical landscape. Protecting data privacy while ensuring security is a delicate balance. The use of surveillance technologies, for instance, can enhance security but also infringe on individual privacy rights. Striking the right balance between security measures and the protection of civil liberties is a critical ethical challenge. Additionally, the development of offensive cyber capabilities, such as malware and hacking tools, raises questions about their use in conflict and the potential for unintended collateral damage.

The ethical implications of autonomous weapons systems are another area of significant concern. These systems, capable of making life-and-death decisions without human intervention, challenge traditional notions of accountability and moral responsibility in warfare. The potential for misuse, accidental escalation of conflicts, and the dehumanization of warfare necessitates robust ethical guidelines and international agreements to govern their development and deployment.

Environmental considerations also play a vital role in the ethical assessment of new technologies. The impact of technological advancements on the environment and future generations must be factored into decision-making processes. Sustainable development and the precautionary principle should guide the deployment of technologies to ensure that they do not exacerbate environmental degradation or contribute to climate change.

In conclusion, the ethical considerations surrounding the use of advanced technologies are multifaceted and complex. Ensuring that these technologies are developed and deployed responsibly requires a commitment to transparency, accountability, and inclusivity. By fostering a culture of ethical awareness and establishing robust regulatory frameworks, society can harness the benefits of technological innovation while safeguarding against potential harms. This delicate balance will be crucial in shaping a future where technology serves the greater good and contributes to a just and equitable world.

Legal Frameworks

The establishment of robust legal frameworks is essential for governing the use of advanced technologies in a manner that promotes safety, fairness, and accountability. As these technologies become increasingly integrated into various aspects of society, the need for comprehensive and adaptive legal systems grows more critical. Legal frameworks provide the necessary guidelines and regulations to ensure that technological advancements are developed and deployed ethically, minimizing potential risks and maximizing benefits.

In the realm of artificial intelligence, legal frameworks must address issues such as data privacy, algorithmic transparency, and accountability. The General Data Protection Regulation (GDPR) in the European Union is a prime example of legislation designed to protect individual privacy rights in the digital age. By setting strict guidelines on data collection, processing, and storage, the GDPR aims to give individuals greater control over their personal information while holding organizations accountable for data breaches and misuse.

The development and deployment of autonomous systems, including self-driving vehicles and drones, also require robust legal frameworks to address safety and liability concerns. Regulations must establish clear standards for testing, certification, and operation of these systems to ensure public safety. Additionally, liability frameworks must be developed to determine responsibility in the event of accidents or malfunctions, balancing the interests of manufacturers, operators, and consumers.

In the field of biotechnology, legal frameworks must navigate complex ethical and safety issues related to genetic engineering and synthetic biology. The Cartagena Protocol on Biosafety is an international agreement that seeks to ensure the safe handling, transport, and use of living modified organisms (LMOs) resulting from modern biotechnology. By establishing guidelines for risk assessment and management, the protocol aims to protect biodiversity and human health from potential adverse effects of biotechnology.

Cybersecurity presents another critical area where legal frameworks are essential. The increasing prevalence of cyber threats requires comprehensive regulations to protect critical infrastructure, secure sensitive data, and ensure national security. Laws such as the Cybersecurity Information Sharing Act (CISA) in the United States encourage the sharing of cybersecurity threat information between the government and private sector, enhancing the collective ability to detect and respond to cyber threats.

International cooperation is vital for the effective implementation of legal frameworks governing advanced technologies. Many of the challenges posed by these technologies transcend national borders, necessitating collaborative efforts to develop harmonized regulations and standards. Treaties, conventions, and international organizations play a crucial role in facilitating this cooperation, ensuring that legal frameworks are consistent and effective globally.

The rapid pace of technological innovation requires that legal frameworks be adaptable and forward-looking. Policymakers must stay informed about emerging technologies and their potential implications to craft regulations that remain relevant and effective. This involves continuous dialogue with industry experts, technologists, and ethicists to understand the evolving landscape and anticipate future challenges.

In conclusion, the development of comprehensive and adaptive legal frameworks is fundamental to ensuring that advanced technologies are used responsibly and ethically. By addressing key issues such as data privacy, safety, liability, and international cooperation, these frameworks provide the necessary structure to manage the risks and benefits of technological innovation. As technology continues to evolve, so too must the legal systems that govern its use, ensuring that progress is achieved in a manner that safeguards the interests of society.

Technological Solutions

Technological solutions play a crucial role in addressing the multifaceted challenges of modern society. These solutions span various fields, from healthcare and environmental sustainability to cybersecurity and artificial intelligence. In healthcare, advanced technologies like telemedicine and wearable devices are revolutionizing patient care, providing real-time monitoring and remote consultations that enhance access to medical services. Innovations in medical imaging and robotic surgery are improving diagnostic accuracy and surgical outcomes, making healthcare more efficient and effective.

Environmental sustainability is another domain where technological solutions are making significant impacts. Renewable energy technologies such as solar, wind, and hydroelectric power are essential in reducing carbon emissions and combating climate change. Innovations in energy storage and smart grid systems are enhancing the efficiency and reliability of renewable energy sources, facilitating their integration into the existing energy infrastructure. Additionally, advancements in waste management technologies, including recycling and biodegradable materials, are helping to address the global waste crisis and promote sustainable practices.

In cybersecurity, the development of advanced encryption methods and intrusion detection systems is critical in protecting sensitive data and infrastructure from cyber threats. Machine learning and artificial intelligence are increasingly being used to detect and respond to cyber-attacks in real time, improving the ability to defend against sophisticated cyber adversaries. Blockchain technology is also gaining traction as a secure method for transaction verification and data integrity, offering robust solutions for digital security.

Artificial intelligence is transforming various industries by enabling more efficient and intelligent decision-making processes. In manufacturing, AI-driven automation and predictive maintenance are optimizing production processes and reducing downtime. In finance, AI algorithms are enhancing fraud detection, risk management, and personalized financial services. The integration of AI in supply chain management is improving logistics, inventory management, and demand forecasting, leading to more efficient and responsive supply chains.

The field of education is also benefiting from technological solutions. E-learning platforms and digital resources are expanding access to education, enabling lifelong learning and skill development. Virtual and augmented reality technologies are creating immersive learning experiences that enhance student engagement and comprehension. Data analytics is being used to tailor educational content to individual learning needs, improving educational outcomes.

In transportation, technological innovations are reshaping mobility and logistics. Autonomous vehicles and electric cars are paving the way for safer, more efficient, and environmentally friendly transportation systems. Smart traffic management systems are reducing congestion and improving urban mobility. Innovations in logistics technologies, such as drone delivery and automated warehouses, are enhancing the efficiency and speed of goods distribution.

Overall, the integration of technological solutions across various sectors is driving significant improvements in efficiency, sustainability, and quality of life. These innovations are not only addressing current challenges but also paving the way for future advancements. By embracing and investing in technological solutions, society can build a more resilient, sustainable, and prosperous future.

Chapter 8: Building Trust and Transparency

Importance of Trust in AI

Public Perception of AI

The public perception of artificial intelligence is a complex and multifaceted issue, shaped by a combination of popular culture, media representation, and actual advancements in technology. AI is often depicted in extremes, either as a groundbreaking innovation that can solve many of humanity's problems or as a potential threat that could lead to job losses, privacy invasions, or even existential risks. These polarized views are influenced by the portrayal of AI in movies, television, and literature, which frequently dramatize both its capabilities and its dangers.

In reality, AI encompasses a wide range of technologies that are already integrated into daily life, from virtual assistants like Siri and Alexa to recommendation algorithms on streaming platforms and social media. These applications have generally been accepted by the public, as they offer convenience and personalized experiences. However, concerns persist about data privacy, the ethics of AI decision-making, and the potential for bias in AI systems. These concerns are not unfounded, as instances of biased algorithms and data breaches have highlighted the need for robust ethical standards and regulatory oversight.

Public opinion is also shaped by the level of understanding people have about AI. There is a significant knowledge gap, with many people not fully understanding how AI works or the extent of its current and potential applications. This gap can lead to misunderstandings and misplaced fears. Efforts to educate the public about AI, including its benefits and limitations, are crucial in fostering a more informed and balanced perspective.

Moreover, trust in AI varies across different demographic groups and regions. Factors such as age, education level, and cultural context can influence how people perceive AI. Younger individuals and those with higher levels of education tend to be more optimistic about the potential of AI, while older individuals and those with less education may be more skeptical or fearful. Cultural differences also play a role, as societies with different historical experiences and values may view the trade-offs of AI technology differently.

The role of policymakers and industry leaders is critical in shaping public perception. Transparent communication about how AI is being used, the measures in place to protect privacy and security, and the ethical guidelines governing AI development can help build public trust. Additionally, involving diverse stakeholders in the development and implementation of AI technologies can ensure that a broad range of perspectives and concerns are considered, leading to more equitable and accepted outcomes.

In summary, the public perception of AI is influenced by a mix of media portrayals, personal experiences, and broader societal factors. Addressing the knowledge gap through education, ensuring ethical and transparent AI practices, and engaging with diverse communities are essential steps in building a more nuanced and positive perception of AI. This approach can help society harness the benefits of AI while mitigating its risks, leading to more informed and balanced views on this transformative technology.

Building Trust through Transparency

Building trust in artificial intelligence systems through transparency is essential for their widespread acceptance and responsible use. Transparency involves clear communication about how AI systems operate, their limitations, and the measures in place to ensure their ethical deployment. This approach can help mitigate fears and misconceptions, fostering a more informed public dialogue about AI technologies.

One of the critical aspects of transparency is the explainability of AI algorithms. People need to understand how decisions are made by these systems, especially in high-stakes areas such as healthcare, finance, and criminal justice. Explainable AI (XAI) aims to make the decision-making processes of AI systems more understandable to non-experts. By providing insights into how AI reaches its conclusions, XAI helps build trust and ensures that users can challenge and question these decisions if necessary.

Another crucial element is the openness about the data used to train AI systems. Data biases can lead to unfair or discriminatory outcomes, which can significantly undermine trust. Ensuring that the datasets are representative and free from biases is essential. Moreover, organizations should be transparent about how they collect, store, and use data, adhering to robust data privacy standards and regulations such as the GDPR.

In addition to technical transparency, ethical transparency is also vital. This involves openly discussing the ethical considerations and potential impacts of AI technologies. By engaging with various stakeholders, including ethicists, policymakers, and the public, developers can address ethical concerns and ensure that AI systems align with societal values.

Public engagement and education are also key components of building trust through transparency. Providing clear and accessible information about AI technologies helps demystify them and reduces the fear of the unknown. Public forums, educational programs, and collaborative initiatives can facilitate a better understanding of AI, encouraging informed public discourse and participation in decision-making processes.

Moreover, regulatory frameworks play a significant role in ensuring transparency. Governments and regulatory bodies need to establish guidelines and standards that mandate transparency in AI development and deployment. These regulations should require organizations to disclose information about their AI systems, including the data used, the algorithms' functioning, and the measures taken to ensure ethical use.

Transparency also extends to addressing the potential risks and failures of AI systems. Organizations should be honest about the limitations and vulnerabilities of their AI technologies and have mechanisms in place to address and mitigate any negative consequences. This includes developing robust protocols for handling AI system failures and ensuring accountability when things go wrong.

In conclusion, building trust in AI through transparency requires a multifaceted approach that includes explainable algorithms, openness about data usage, ethical considerations, public engagement, and regulatory oversight. By fostering a culture of transparency, organizations can enhance public confidence in AI technologies, ensuring their responsible and beneficial integration into society.

Case Studies in Trust-building

The process of building trust in artificial intelligence technologies can be best understood through various case studies where transparency, accountability, and stakeholder engagement have been successfully implemented. One notable example is the collaboration between the Royal Free London NHS Foundation Trust and DeepMind, an AI subsidiary of Alphabet Inc. This partnership aimed to improve patient care through the development of an AI-based application called Streams, designed to alert clinicians to signs of acute kidney injury.

Initially, the project faced significant public backlash due to concerns over patient data privacy. Critics argued that the data-sharing agreement between the NHS and DeepMind lacked adequate transparency and did not sufficiently inform patients about how their data would be used. In response to these concerns, the partnership made several changes to enhance transparency and public trust. They conducted independent audits, published detailed reports on data usage, and engaged with patient advocacy groups to address privacy issues and ensure compliance with data protection regulations. These steps helped rebuild public trust and demonstrated the importance of clear communication and accountability in the use of AI in healthcare.

Another case study that illustrates effective trust-building through transparency is IBM's Watson for Oncology. Watson for Oncology is an AI system designed to assist oncologists in diagnosing and treating cancer. To ensure the system's recommendations were trusted by medical professionals, IBM implemented a transparent approach to the development and deployment of Watson. They provided detailed documentation on the data sources and algorithms used, as well as the training process for the AI system. Additionally, IBM collaborated with leading cancer institutes and involved oncologists in the continuous evaluation and improvement of Watson's performance. By fostering an environment of collaboration and openness, IBM was able to build confidence in the AI system's capabilities among healthcare providers.

In the financial sector, the case of the Dutch bank ING offers insights into trust-building through responsible AI practices. ING developed an AI-driven system to enhance customer service and risk management. Recognizing the potential risks associated with AI, ING established an ethical framework that included principles such as fairness, accountability, and transparency. The bank also created an AI governance board to oversee the development and implementation of AI systems, ensuring they adhered to ethical standards and regulatory requirements. By proactively addressing ethical concerns and maintaining transparency in their AI practices, ING successfully gained the trust of its customers and regulators.

These case studies highlight the importance of transparency, stakeholder engagement, and ethical governance in building trust in AI technologies. By openly communicating about how AI systems are developed and used, involving stakeholders in the process, and adhering to ethical standards, organizations can mitigate fears and concerns, fostering a more positive perception of AI. These examples serve as valuable lessons for other organizations seeking to implement AI in a manner that is both responsible and trusted by the public.

Transparent AI Development

Open Source AI

The concept of open-source artificial intelligence is transforming the landscape of technology development, fostering a collaborative environment where researchers, developers, and organizations can share knowledge and resources. Open-source AI refers to AI projects whose source code is made available to the public, allowing anyone to inspect, modify, and enhance the software. This approach has numerous benefits, including accelerating innovation, improving transparency, and democratizing access to cutting-edge technologies.

One of the key advantages of open-source AI is the acceleration of innovation. By making AI tools and frameworks available to a global community, open-source projects enable a diverse group of contributors to collaborate, share ideas, and build upon each other's work. This collective effort often leads to more rapid advancements and the development of robust, well-tested solutions. Prominent examples of open-source AI projects include TensorFlow, developed by Google, and PyTorch, developed by Facebook. These frameworks have become foundational tools for AI research and development, providing essential building blocks for creating complex machine learning models.

Transparency is another significant benefit of open-source AI. When the source code of AI systems is publicly available, it allows for greater scrutiny and validation by the community. Researchers and practitioners can examine the underlying algorithms, identify potential biases, and ensure that the systems are performing as intended. This level of transparency is crucial for building trust in AI technologies, particularly in applications where fairness and accountability are paramount, such as healthcare, finance, and criminal justice.

Open-source AI also democratizes access to advanced technologies. Traditionally, cutting-edge AI research and development have been concentrated within large corporations and well-funded institutions. However, open-source initiatives lower the barriers to entry, enabling smaller organizations, startups, and individual developers to access and leverage powerful AI tools. This democratization fosters a more inclusive ecosystem where a broader range of voices and perspectives can contribute to the development of AI technologies.

Moreover, open-source AI promotes education and skill development. By providing access to state-of-the-art tools and real-world projects, open-source platforms offer valuable learning opportunities for students and professionals. Aspiring AI practitioners can experiment with open-source frameworks, participate in collaborative projects, and gain hands-on experience that is essential for developing expertise in the field.

Despite its numerous benefits, open-source AI also presents certain challenges. The quality and security of open-source projects can vary, and there is a need for robust governance and maintenance to ensure that these projects remain reliable and secure. Additionally, the open nature of these projects can sometimes lead to fragmentation, where multiple versions or forks of a project evolve independently, potentially diluting the impact of the collective effort.

In conclusion, open-source AI is a powerful paradigm that enhances innovation, transparency, and accessibility in the field of artificial intelligence. By fostering collaboration and democratizing access to advanced tools, open-source AI initiatives are driving significant advancements and enabling a wider range of contributors to shape the future of AI technology. As the open-source community continues to grow and evolve, it will play a crucial role in addressing the challenges and maximizing the benefits of AI for society.

Transparency in AI Algorithms

Transparency in AI algorithms is a crucial factor in fostering trust, ensuring ethical use, and promoting fairness in AI applications. Transparency refers to the ability to understand and interpret the workings and decisions of AI systems. This involves making the processes by which algorithms function clear and comprehensible to users, developers, and regulators.

One of the key aspects of transparency is explainability, which involves providing clear, understandable explanations of how AI systems reach their decisions. Explainable AI (XAI) is essential in high-stakes areas such as healthcare, finance, and criminal justice, where understanding the rationale behind decisions can have significant implications. For instance, in healthcare, it is vital for medical professionals to comprehend how an AI system arrived at a particular diagnosis or treatment recommendation. This not only ensures accountability but also enables medical practitioners to trust and effectively use AI tools in their decision-making processes.

Transparency also involves the disclosure of the data used to train AI models. The quality and bias of data can significantly impact the outcomes of AI systems. By making the training data available and subject to scrutiny, developers can ensure that the data is representative and free from biases that could lead to unfair or discriminatory results. Open datasets and transparent data practices contribute to the fairness and reliability of AI systems.

Moreover, transparency requires that the algorithms themselves are open to inspection. Open-source AI frameworks and models allow developers and researchers to examine the underlying code and logic, identify potential flaws or biases, and contribute to improving the systems. This collaborative approach not only enhances the robustness of AI technologies but also fosters a sense of community and shared responsibility in the AI field.

Transparency in AI also extends to the development and implementation processes. Organizations must be open about their AI development practices, including the goals, methodologies, and potential limitations of their systems. This involves regular audits, impact assessments, and the publication of findings related to the performance and ethical implications of AI technologies. By being transparent about the processes and considerations involved in AI development, organizations can build trust with stakeholders and the broader public.

Regulatory frameworks play a crucial role in promoting transparency in AI. Governments and regulatory bodies need to establish guidelines and standards that mandate transparency in the design, development, and deployment of AI systems. These regulations should require organizations to disclose relevant information about their AI technologies, including how they function, the data they use, and the measures taken to ensure ethical use.

In conclusion, transparency in AI algorithms is fundamental to ensuring that AI technologies are trusted, ethical, and fair. By promoting explainability, disclosing data sources, opening algorithms to inspection, and adhering to transparent development practices, organizations can build trust and accountability in AI systems. Regulatory frameworks further reinforce the importance of transparency, ensuring that AI technologies are developed and used in ways that benefit society as a whole. This approach not only enhances the credibility of AI but also encourages broader acceptance and responsible use of these transformative technologies.

Regulatory Standards

The establishment of robust regulatory standards for artificial intelligence is crucial to ensure the ethical and safe deployment of AI technologies. Regulatory frameworks provide the necessary guidelines and oversight to address the potential risks and challenges associated with AI, promoting transparency, accountability, and fairness.

One of the primary goals of regulatory standards is to protect data privacy. As AI systems often rely on large datasets to function effectively, regulations such as the General Data Protection Regulation (GDPR) in the European Union set stringent requirements for data collection, storage, and usage. The GDPR mandates that organizations obtain explicit consent from individuals before collecting their data and ensure that the data is used transparently and securely. Such regulations are vital in maintaining public trust and protecting individuals' privacy rights.

Another important aspect of AI regulation is ensuring the fairness and accountability of AI algorithms. Bias in AI systems can lead to unfair and discriminatory outcomes, particularly in sensitive areas like hiring, lending, and law enforcement. Regulatory standards can require organizations to conduct regular audits of their AI systems to identify and mitigate biases. Additionally, transparency requirements can compel companies to disclose the methodologies and datasets used in their AI models, allowing for external scrutiny and validation.

The development of AI-specific ethical guidelines is also a key component of regulatory standards. These guidelines can address issues such as the ethical implications of autonomous decision-making, the potential impact of AI on employment, and the societal consequences of widespread AI adoption. By providing a clear ethical framework, regulations can guide the responsible development and deployment of AI technologies.

International cooperation is essential for the effective regulation of AI. As AI technologies are developed and deployed globally, harmonized regulatory standards can help ensure consistency and prevent regulatory arbitrage. Organizations such as the Organisation for Economic Co-operation and Development (OECD) and the International Telecommunication Union (ITU) play a critical role in facilitating international dialogue and developing global standards for AI.

Moreover, regulatory standards need to be adaptive and forward-looking to keep pace with the rapid advancements in AI technology. Policymakers must stay informed about emerging trends and potential risks to craft regulations that remain relevant and effective. This requires continuous engagement with AI researchers, industry experts, and other stakeholders to understand the evolving landscape and anticipate future challenges.

In conclusion, the establishment of comprehensive regulatory standards is fundamental to ensuring the ethical, transparent, and accountable use of artificial intelligence. By protecting data privacy, ensuring fairness, promoting ethical guidelines, and fostering international cooperation, regulatory frameworks can mitigate the risks associated with AI and maximize its benefits for society. Adaptive and forward-looking regulations will be crucial in addressing the dynamic and complex nature of AI, ensuring that its development and deployment align with societal values and public interest.

Ethical AI Communication

Clear Communication of AI Decisions

The clear communication of AI decisions is a fundamental aspect of building trust and ensuring the ethical deployment of artificial intelligence systems. Transparency in how AI decisions are made helps users understand the rationale behind these decisions, fostering confidence in the technology. Explainable AI, a branch of AI focused on making machine learning models more interpretable, plays a crucial role in this regard. By providing insights into the inner workings of algorithms, explainable AI allows users to see how data is processed and decisions are derived.

One critical area where clear communication of AI decisions is essential is healthcare. When AI systems assist in diagnosing diseases or recommending treatments, medical professionals need to understand how these conclusions are reached. This understanding ensures that AI tools can be used effectively and safely in clinical settings. For example, an AI system that analyzes medical images to detect tumors must be able to explain its findings so that doctors can verify the results and make informed decisions about patient care.

In the financial sector, transparency in AI decision-making helps prevent bias and discrimination in lending and credit scoring. Financial institutions use AI to evaluate loan applications and assess credit risk, and it is vital that these decisions are made fairly and transparently. By providing clear explanations for why certain applications are approved or denied, AI systems can help ensure that lending practices are equitable and comply with regulatory standards.

Moreover, clear communication of AI decisions is crucial in law enforcement and criminal justice. AI algorithms are increasingly used to predict criminal behavior, assess the risk of reoffending, and assist in sentencing decisions. These applications have significant implications for individuals' lives, and it is imperative that the AI systems used are transparent and their decision-making processes are understandable. This transparency can help safeguard against wrongful convictions and ensure that justice is administered fairly.

Achieving clear communication of AI decisions involves several strategies. One approach is the use of visual aids and interactive tools that allow users to explore how AI models work and how they arrive at specific decisions. These tools can demystify complex algorithms and make the decision-making process more accessible to non-experts. Additionally, AI developers can provide detailed documentation and user guides that explain the logic and assumptions behind their models.

Another important aspect is involving diverse stakeholders in the development and evaluation of AI systems. By engaging ethicists, legal experts, and representatives from affected communities, AI developers can ensure that their systems are designed with fairness and transparency in mind. This collaborative approach helps identify potential biases and ethical issues early in the development process, leading to more trustworthy AI systems.

In conclusion, the clear communication of AI decisions is essential for building trust, ensuring ethical use, and promoting fairness across various sectors. Explainable AI, transparent practices, and stakeholder engagement are key strategies in achieving this goal. By making AI decision-making processes understandable and accessible, we can harness the benefits of AI while mitigating its risks, ultimately leading to more equitable and trustworthy AI applications.

Human-AI Interaction

The interaction between humans and artificial intelligence systems is a critical area of study that shapes how effectively and ethically AI technologies can be integrated into society. Effective human-AI interaction depends on several factors, including user interface design, the transparency of AI decision-making processes, and the ability of AI systems to understand and respond to human inputs accurately and empathetically.

One of the primary goals in designing human-AI interactions is to create interfaces that are intuitive and user-friendly. This involves understanding the user's needs and preferences and designing systems that can adapt to various levels of technical proficiency. For example, voice-activated assistants like Siri and Alexa have become widely popular due to their ability to understand natural language and provide relevant responses, making technology accessible to a broader audience. The simplicity of these interactions hides the complex algorithms at work, ensuring that users can engage with the AI without needing to understand its technical underpinnings.

Another crucial aspect of human-AI interaction is the transparency of AI systems. Users need to trust that AI systems are making decisions fairly and accurately. This trust can be fostered by providing clear explanations of how AI systems reach their conclusions. Explainable AI is an emerging field focused on developing methods to make AI decisions understandable to humans. By offering insights into the decision-making processes, these systems can help users feel more confident in the outcomes provided by AI, particularly in high-stakes areas like healthcare and finance.

Empathy and contextual understanding are also vital in human-AI interaction. AI systems must be able to interpret human emotions and context to respond appropriately. This capability is especially important in applications like customer service, mental health support, and education. For instance, AI chatbots used in customer service are designed to recognize frustration or confusion in a user's language and respond in a manner that is calming and helpful. Similarly, educational AI tools can adapt to a student's learning pace and style, providing personalized support that enhances the learning experience.

Moreover, the ethical considerations surrounding human-AI interaction cannot be overlooked. AI systems must be designed to respect user privacy and autonomy. Ensuring that AI interactions are consensual and that users have control over how their data is used is crucial in maintaining ethical standards. Regulatory frameworks and ethical guidelines play an essential role in this context, providing a basis for developing and implementing AI technologies that prioritize user rights and societal well-being.

In conclusion, the effective and ethical integration of AI into daily life depends on thoughtful design and transparency in human-AI interactions. By focusing on user-friendly interfaces, transparent decision-making processes, and empathetic, context-aware responses, developers can create AI systems that enhance human capabilities and foster trust. As AI continues to evolve, ongoing research and dialogue about the ethical implications and best practices for human-AI interaction will be essential in ensuring that these technologies benefit society as a whole.

Case Studies in Ethical Communication

Ethical communication in artificial intelligence involves ensuring transparency, accountability, and fairness in how AI systems operate and interact with users. Examining various case studies helps illustrate how these principles can be effectively implemented in real-world scenarios.

One significant example is the implementation of AI systems in the criminal justice system by COMPAS (Correctional Offender Management Profiling for Alternative Sanctions). This AI tool was designed to assess the risk of recidivism among defendants. However, its use raised ethical concerns due to allegations of bias against African American defendants. The lack of transparency about how COMPAS reached its decisions and the proprietary nature of its algorithms fueled criticism and legal challenges. This case underscores the importance of transparency and the need for explainable AI systems in maintaining ethical standards and public trust.

Another instructive case is the use of AI in healthcare by IBM's Watson for Oncology. Designed to assist oncologists in diagnosing and treating cancer, Watson for Oncology exemplifies the need for clear communication and transparency. IBM made efforts to document the data sources and training methodologies used to develop Watson, involving oncologists in the evaluation process to ensure the system's recommendations were accurate and trustworthy. This collaboration between technologists and healthcare professionals helped build trust and demonstrated the potential benefits of AI when ethical considerations are prioritized.

A third case study involves Google's DeepMind and its partnership with the UK's National Health Service (NHS) to develop an app called Streams, which aimed to improve patient care by detecting acute kidney injuries. Initially, the project faced significant backlash due to privacy concerns and a lack of transparency about how patient data was being used. In response, DeepMind and the NHS increased their efforts to engage with the public and patient advocacy groups, conducting independent audits and providing detailed reports on data usage. This approach helped rebuild trust and highlighted the importance of ethical communication and transparency in managing sensitive data.

These case studies illustrate that ethical communication in AI involves not only transparency in decision-making processes but also proactive engagement with stakeholders. Ensuring that AI systems are explainable, fair, and accountable requires continuous collaboration between developers, users, and regulators. By learning from these examples, future AI implementations can better address ethical challenges and foster greater public trust in AI technologies.

Chapter 9: The Role of AI in Society

Enhancing Human Capabilities

AI in Healthcare

Artificial intelligence is revolutionizing healthcare by enhancing diagnostic accuracy, improving treatment plans, and increasing the efficiency of administrative tasks. AI-driven tools are capable of analyzing vast amounts of medical data to identify patterns and predict outcomes. For example, AI algorithms can interpret medical images, such as X-rays and MRIs, with a level of accuracy comparable to human radiologists, aiding in the early detection of diseases like cancer. These systems use deep learning techniques to continuously improve their diagnostic capabilities.

AI is also transforming personalized medicine. By analyzing genetic information, lifestyle data, and clinical records, AI systems can tailor treatment plans to individual patients, optimizing outcomes and minimizing adverse effects. This personalized approach is particularly beneficial in the treatment of complex conditions like cancer, where treatment efficacy can vary widely among patients.

Moreover, AI is streamlining administrative processes in healthcare. Natural language processing (NLP) enables the automation of routine tasks such as patient record management, appointment scheduling, and billing. This automation not only reduces the workload for healthcare professionals but also minimizes the risk of errors, ensuring more accurate and efficient service delivery.

Another significant application of AI in healthcare is in predictive analytics. AI models can predict disease outbreaks, patient admissions, and potential complications by analyzing historical data and current trends. These predictive capabilities enable healthcare providers to allocate resources more effectively and implement preventive measures, improving overall patient care and system efficiency.

However, the integration of AI in healthcare also raises ethical and regulatory challenges. Ensuring the privacy and security of patient data is paramount, as is addressing potential biases in AI algorithms that could lead to disparities in care. Transparent and accountable AI development practices are essential to build trust among healthcare providers and patients.

In summary, AI is playing a transformative role in healthcare by enhancing diagnostics, personalizing treatment, improving administrative efficiency, and enabling predictive analytics. While the benefits are significant, addressing the ethical and regulatory challenges is crucial to ensuring the responsible and equitable implementation of AI technologies in healthcare.

AI in Education

The integration of artificial intelligence in education is transforming traditional learning paradigms, offering personalized learning experiences, and enhancing administrative efficiencies. AI-driven educational tools are capable of tailoring learning experiences to individual students' needs, identifying strengths and weaknesses, and adapting content accordingly. This personalized approach ensures that each student can progress at their own pace, receiving the support and challenges appropriate to their level.

One significant application of AI in education is the use of intelligent tutoring systems (ITS). These systems provide one-on-one tutoring tailored to the student's learning style and pace. By analyzing the student's interactions, the ITS can identify areas where the student struggles and adjust the instruction to address these gaps. This personalized attention helps to ensure that students do not fall behind and can master the material effectively.

In addition to personalized learning, AI is enhancing the assessment process. Automated grading systems can efficiently evaluate large volumes of assignments and exams, providing immediate feedback to students. This instant feedback loop helps students understand their mistakes and learn from them in real time. Moreover, AI can analyze patterns in student performance to provide educators with insights into common areas of difficulty, allowing for targeted instructional interventions.

Administrative tasks in educational institutions also benefit from AI. Enrollment processes, scheduling, and resource allocation can be streamlined through AI-powered systems, reducing the administrative burden on staff and allowing them to focus more on student engagement and support. These systems can also help in identifying at-risk students early by analyzing data such as attendance, grades, and engagement levels, enabling timely interventions to support student success.

Furthermore, AI fosters an inclusive learning environment by providing support for students with special needs. For example, speech recognition and natural language processing technologies can assist students with disabilities, offering them tools to participate fully in classroom activities. AI can also translate educational materials into multiple languages, making learning more accessible to non-native speakers.

However, the implementation of AI in education must be approached with careful consideration of ethical and privacy concerns. Ensuring the protection of student data is paramount, and transparency in how AI systems operate and make decisions is critical to maintaining trust. Educators and policymakers must work together to establish guidelines and regulations that safeguard student information and promote the ethical use of AI technologies.

In conclusion, AI is revolutionizing education by offering personalized learning, enhancing assessments, and streamlining administrative tasks. These advancements create more effective and inclusive educational environments, ultimately improving student outcomes. As AI continues to evolve, it holds the potential to further transform education, making it more adaptive, efficient, and accessible for all learners.

AI in Workforce Development

Artificial intelligence is significantly enhancing workforce development by providing tools that improve training, skill acquisition, and job matching. AI-driven platforms can analyze a vast array of data to identify skills gaps and recommend personalized learning paths for individuals. These platforms utilize machine learning algorithms to adapt to the learner's progress, ensuring that training is tailored to their specific needs and pace.

One prominent application of AI in workforce development is in the realm of personalized training programs. AI systems can create customized training modules that address the unique strengths and weaknesses of each employee. By continuously analyzing performance data, these systems can adjust the difficulty and focus of the training materials to ensure optimal learning outcomes. This personalized approach not only enhances the efficiency of the training process but also improves retention and application of new skills.

Moreover, AI facilitates more effective job matching by analyzing the skills, experiences, and preferences of job seekers against the requirements of available positions. This results in better-aligned job placements, which benefit both employers and employees by increasing job satisfaction and reducing turnover rates. AI-powered platforms can also predict future skills needs based on industry trends, helping workers stay ahead in their careers by acquiring relevant skills before they become critical.

AI is also revolutionizing the recruitment process. Automated systems can sift through large volumes of applications to identify the best candidates, reducing the time and cost associated with hiring. These systems can evaluate a wide range of factors, including work history, education, and even behavioral traits, to determine the best fit for a role. This enables recruiters to focus on more strategic tasks, such as interviewing and assessing cultural fit.

Furthermore, AI-driven analytics provide valuable insights into workforce dynamics. Organizations can use these insights to make informed decisions about talent management, such as identifying high-potential employees, planning succession strategies, and designing effective employee engagement initiatives. By leveraging AI, companies can create a more agile and responsive workforce that is better equipped to adapt to changing market conditions.

In addition to these practical applications, AI also plays a crucial role in fostering a culture of continuous learning and development. By making learning resources easily accessible and tailoring them to individual needs, AI encourages employees to take ownership of their professional growth. This culture of continuous improvement is essential for organizations looking to remain competitive in a rapidly evolving business landscape.

However, the integration of AI in workforce development must be managed carefully to address ethical considerations and ensure fairness. It is crucial to ensure that AI systems are free from biases that could perpetuate inequalities. Transparent algorithms and regular audits can help maintain the integrity of AI-driven processes and build trust among users.

In summary, AI is transforming workforce development by personalizing training, enhancing job matching, streamlining recruitment, and providing valuable workforce analytics. These advancements not only improve organizational efficiency but also empower employees to achieve their full potential. As AI technology continues to evolve, its impact on workforce development will likely grow, driving further innovations and improvements in how we train and manage talent.

Societal Impacts

AI and Employment

Artificial intelligence is reshaping the employment landscape, creating new opportunities while also presenting challenges. One of the key impacts of AI on employment is the automation of routine tasks. Automation has the potential to increase efficiency and reduce costs in various industries, from manufacturing to customer service. For instance, AI-powered robots can perform repetitive tasks on assembly lines, while chatbots can handle customer inquiries, freeing human workers to focus on more complex and creative tasks.

AI is also driving the demand for new skills. As automation takes over certain job functions, there is a growing need for workers with expertise in AI development, data analysis, and machine learning. Educational institutions and training programs are increasingly focusing on equipping individuals with these skills to prepare them for the evolving job market. This shift necessitates continuous learning and adaptation, as the rapid pace of technological advancements requires workers to stay current with the latest developments.

Moreover, AI facilitates better job matching and recruitment processes. AI algorithms can analyze vast amounts of data to match job seekers with positions that best fit their skills and experiences, thereby improving job placement efficiency. Recruitment platforms powered by AI can screen resumes and conduct initial interviews, enabling recruiters to focus on candidates who are the best fit for the role. This not only streamlines the hiring process but also reduces biases that can occur in traditional recruitment methods.

However, the integration of AI in employment also raises significant concerns. One major issue is the potential displacement of workers due to automation. While AI creates new job opportunities, it can also render certain roles obsolete. This displacement disproportionately affects low-skilled workers who may not have the resources or opportunities to reskill. Addressing this challenge requires collaborative efforts from governments, businesses, and educational institutions to provide retraining and support for displaced workers.

Additionally, ethical considerations in AI deployment are crucial. Ensuring that AI systems are designed and used in ways that are fair, transparent, and accountable is essential to maintaining public trust. For example, AI algorithms used in hiring must be free from biases that could disadvantage certain groups.

Transparency in how these algorithms make decisions and accountability mechanisms for addressing any adverse impacts are necessary to uphold ethical standards.

In conclusion, AI is significantly transforming employment by automating routine tasks, creating new skill demands, and enhancing recruitment processes. While it offers numerous benefits, it also poses challenges such as worker displacement and ethical concerns. Addressing these challenges requires a multifaceted approach involving continuous learning, supportive policies, and ethical AI practices. By navigating these complexities, society can harness the potential of AI to create a more efficient and equitable job market.

AI and Economic Growth

Artificial intelligence is a driving force behind economic growth, transforming industries and creating new opportunities for innovation and efficiency. AI technologies enhance productivity by automating routine tasks, optimizing supply chains, and enabling more effective decision-making processes. For example, in manufacturing, AI-driven automation improves production rates and reduces errors, leading to higher output and lower costs.

The financial sector also benefits significantly from AI. Advanced algorithms can analyze vast datasets to identify trends and make predictions, improving investment strategies and risk management. AI systems in finance can process transactions faster and more accurately than humans, enhancing operational efficiency and customer satisfaction.

Moreover, AI stimulates economic growth by fostering innovation. Startups and established companies alike are leveraging AI to develop new products and services, from personalized marketing solutions to advanced healthcare diagnostics. These innovations not only create new markets but also drive competition, encouraging further advancements and investments in AI technology.

In the healthcare industry, AI contributes to economic growth by improving patient outcomes and reducing costs. AI-powered diagnostic tools can detect diseases at earlier stages, leading to more effective treatments and shorter hospital stays. Additionally, AI can streamline administrative tasks in healthcare, such as scheduling and billing, freeing up resources for patient care.

AI also plays a crucial role in enhancing customer experiences across various sectors. Retail businesses use AI to personalize shopping experiences, recommend products, and optimize inventory management. This personalization increases customer satisfaction and loyalty, driving sales and growth.

However, the integration of AI into the economy comes with challenges that must be addressed. The displacement of workers due to automation is a significant concern. As AI takes over routine tasks, there is a need for policies and programs that support workforce retraining and upskilling. Ensuring that workers can transition to new roles within the evolving job market is essential for maintaining economic stability and growth.

Another challenge is the ethical use of AI. Transparent and accountable AI systems are necessary to build public trust and avoid potential misuse. Regulations and guidelines must be established to ensure that AI technologies are developed and deployed responsibly, with considerations for privacy, fairness, and security.

In conclusion, AI is a powerful catalyst for economic growth, driving productivity, innovation, and improved customer experiences. While it offers significant benefits, it is crucial to address the challenges associated with AI integration to ensure sustainable and inclusive growth. By fostering a supportive environment for AI development and addressing ethical concerns, societies can harness the full potential of AI to drive economic progress.

AI in Social Services

Artificial intelligence is increasingly being utilized in social services to enhance the delivery of support and improve outcomes for individuals and communities. AI-driven tools are transforming how social services are provided, making processes more efficient and enabling more personalized care. These technologies can analyze large datasets to identify trends and patterns, helping social workers to make informed decisions and allocate resources more effectively.

For instance, predictive analytics can be used to identify individuals or families at risk of adverse outcomes, such as homelessness or child neglect, allowing for early intervention. By analyzing historical data and current indicators, AI systems can flag cases that require urgent attention, thereby preventing crises before they occur. This proactive approach not only improves the quality of care but also reduces long-term costs associated with reactive measures.

AI also plays a significant role in streamlining administrative tasks within social services. Natural language processing (NLP) can automate the processing of case notes and documentation, freeing up social workers to spend more time directly engaging with clients. Additionally, AI-powered chatbots can provide 24/7 support and information to individuals seeking assistance, ensuring that help is available at any time.

Moreover, AI can facilitate better coordination among different service providers. By integrating data from various sources, AI systems can create comprehensive profiles of individuals, ensuring that all relevant information is available to social workers. This holistic view enables more coordinated and effective interventions, as social workers can see the full picture of a client's needs and circumstances.

Ethical considerations are paramount in the deployment of AI in social services. Ensuring the privacy and security of sensitive data is crucial, as is addressing potential biases in AI algorithms that could lead to unfair treatment. Transparent AI practices and regular audits are essential to maintain trust and ensure that these technologies are used responsibly.

In conclusion, AI is revolutionizing social services by enhancing predictive capabilities, streamlining administrative tasks, and improving coordination among service providers. While the benefits are substantial, it is essential to address ethical challenges to ensure that AI is used in a manner that respects privacy and promotes fairness. Through thoughtful implementation, AI can significantly improve the effectiveness and efficiency of social services, leading to better outcomes for individuals and communities.

Ethical Considerations

Equity and Inclusion

Ensuring equity and inclusion in the deployment of artificial intelligence technologies is crucial for creating a fair and just society. AI systems have the potential to significantly impact various aspects of life, from healthcare and education to employment and social services. However, without careful design and implementation, these systems can perpetuate existing biases and exacerbate inequalities.

One of the primary challenges in achieving equity and inclusion in AI is addressing bias in data and algorithms. AI systems learn from the data they are trained on, and if this data reflects historical inequalities or biases, the AI can replicate and even amplify these issues. For example, in hiring processes, if an AI system is trained on biased historical data, it may unfairly favor certain demographic groups over others. To counter this, it is essential to use diverse and representative datasets and to implement robust mechanisms for detecting and mitigating bias in AI models.

Another key aspect is ensuring that AI technologies are accessible to all segments of society. This involves making AI tools and resources available to underserved and marginalized communities, providing them with the knowledge and skills needed to utilize these technologies effectively. Educational initiatives and community outreach programs can play a significant role in democratizing access to AI, ensuring that everyone has the opportunity to benefit from its advancements.

Moreover, transparency and accountability are fundamental to promoting equity and inclusion in AI. Organizations developing and deploying AI systems must be transparent about their methodologies, data sources, and decision-making processes. This transparency allows for external scrutiny and validation, helping to build trust and ensure that AI systems operate fairly. Additionally, establishing clear accountability mechanisms ensures that those responsible for AI systems are held accountable for any adverse impacts, fostering a culture of responsibility and ethical conduct.

In the context of social services, AI can be leveraged to enhance the delivery of support to vulnerable populations. For instance, AI-powered systems can identify individuals or families at risk of experiencing poverty, homelessness, or health crises, enabling timely and targeted interventions. However, it is crucial to ensure that these systems are designed and implemented with a focus on equity, ensuring that they do not inadvertently discriminate against or disadvantage any group.

Collaboration across sectors is also vital for advancing equity and inclusion in AI. Governments, businesses, educational institutions, and civil society organizations must work together to develop and enforce standards and best practices for ethical AI. This collaborative approach helps to create a more inclusive AI ecosystem, where diverse perspectives are considered and valued.

In conclusion, promoting equity and inclusion in AI requires a multifaceted approach that addresses bias, ensures accessibility, fosters transparency, and encourages collaboration. By prioritizing these principles, society can harness the transformative potential of AI to create a more equitable and inclusive future. The commitment to ethical AI practices not only enhances the technology's benefits but also ensures that it serves as a force for good in society.

Bias and Fairness

Bias in artificial intelligence is a significant concern that can lead to unfair outcomes, perpetuating existing inequalities and creating new ones. AI systems learn from data, and if this data is biased, the resulting algorithms will likely reflect those biases. For example, if an AI model is trained on hiring data that historically favors one demographic over another, it might continue to favor that demographic in its hiring recommendations.

Addressing bias in AI requires a multi-faceted approach. Firstly, it is essential to use diverse and representative datasets in training AI models. This ensures that the AI system is exposed to a wide range of scenarios and can make fair decisions across different groups. Secondly, developers must implement techniques for detecting and mitigating bias within algorithms. This involves regularly auditing AI systems and adjusting them to correct any biases that are identified.

Transparency in AI processes is also crucial. When organizations are open about how their AI systems work and the data they use, it allows for external scrutiny and helps build trust. Clear documentation and explainable AI models enable stakeholders to understand and challenge the decisions made by AI, ensuring accountability.

Moreover, it is vital to involve diverse teams in the development of AI technologies. Teams with varied backgrounds and perspectives are more likely to recognize and address potential biases that homogeneous teams might overlook. This diversity in development teams can lead to more balanced and fair AI systems.

In practice, ensuring fairness in AI means continuously monitoring and improving these systems. This ongoing effort is necessary to adapt to new data and changing societal norms. Organizations must be committed to ethical AI practices and prioritize fairness in their AI strategies.

In conclusion, addressing bias and ensuring fairness in AI involves using diverse datasets, implementing bias detection techniques, promoting transparency, and involving diverse development teams. These efforts are essential for creating AI systems that are fair, accountable, and beneficial for all members of society. Through diligent and thoughtful approaches, the AI community can work towards minimizing bias and fostering equity in AI applications.

Case Studies in Ethical AI Deployment

The ethical deployment of artificial intelligence is a subject of critical importance as it determines the impact of AI technologies on society. Examining case studies in this domain provides valuable insights into best practices and the challenges involved. One notable example is IBM's Watson for Oncology, an AI system designed to assist oncologists in diagnosing and treating cancer. IBM made substantial efforts to document the sources of data and the methodologies used to train Watson. By involving oncologists in the evaluation process, IBM ensured that the AI's recommendations were reliable and clinically relevant. This transparency and collaboration helped build trust among healthcare professionals, showcasing the importance of stakeholder engagement and clear communication in AI deployment.

Another case study involves the use of AI by the Dutch bank ING, which developed an AI-driven system to enhance customer service and risk management. Recognizing the potential risks associated with AI, ING established an ethical framework that included principles of fairness, accountability, and transparency. The bank also created an AI governance board to oversee the development and implementation of AI systems, ensuring adherence to ethical standards and regulatory requirements. This approach demonstrated how organizations can effectively integrate AI while maintaining ethical integrity.

A third example is Google's DeepMind and its partnership with the UK's National Health Service (NHS) to develop an app called Streams, aimed at improving patient care by detecting acute kidney injuries. Initially, the project faced significant criticism due to privacy concerns and a lack of transparency regarding data usage. In response, DeepMind and the NHS increased their efforts to engage with the public and patient advocacy groups, conducting independent audits and publishing detailed reports on data practices. This proactive stance helped rebuild trust and underscored the necessity of transparency and public engagement in the ethical deployment of AI.

These case studies illustrate the critical elements of ethical AI deployment: transparency, stakeholder engagement, and adherence to ethical principles. They highlight the importance of involving diverse teams in the development process, ensuring that AI systems are designed to be fair, accountable, and beneficial for all users. By following these practices, organizations can harness the power of AI while mitigating risks and fostering public trust. Through thoughtful and responsible approaches, the AI community can set standards that promote the ethical use of AI technologies, ultimately contributing to a more equitable and just society.

Chapter 10: Future Directions in AI and Robotics

Emerging Technologies

Advances in Machine Learning

Advances in machine learning have propelled the field of artificial intelligence into new frontiers, offering transformative potential across various industries. Machine learning, a subset of AI, involves the development of algorithms that enable computers to learn from and make decisions based on data. One significant breakthrough in this area is deep learning, which uses neural networks with many layers to analyze complex patterns in large datasets. This technique has dramatically improved the accuracy and efficiency of tasks such as image and speech recognition.

Reinforcement learning is another critical advancement. In this paradigm, algorithms learn to make sequences of decisions by receiving rewards or penalties, akin to training a dog with treats. This approach has been successfully applied to develop AI systems that can master complex games like Go and StarCraft, outperforming human champions.

Transfer learning has also gained prominence, allowing models trained on one task to be adapted for related tasks with minimal retraining. This capability significantly reduces the time and resources required to develop AI solutions for new applications.

Generative adversarial networks (GANs) represent another innovative leap. GANs consist of two neural networks, a generator and a discriminator, that work together to produce highly realistic synthetic data, such as images and videos. This technology is not only revolutionizing creative industries but also enhancing data augmentation for training other AI models.

The practical applications of these machine learning advancements are vast. In healthcare, AI is aiding in the early detection of diseases through the analysis of medical images and patient data. In finance, AI algorithms are improving fraud detection and optimizing investment strategies. Autonomous vehicles rely on machine learning to navigate and make real-time decisions, paving the way for safer and more efficient transportation systems.

Despite these remarkable advancements, challenges remain. Ensuring the ethical use of AI and mitigating biases in machine learning models are critical areas of ongoing research. Transparency and explainability in AI decisions are essential to build trust and accountability.

In summary, the advances in machine learning, particularly in deep learning, reinforcement learning, transfer learning, and GANs, are driving significant innovations across various sectors. These technologies hold the promise of solving complex problems and enhancing human capabilities, marking an exciting era in the evolution of artificial intelligence.

Robotics Innovations

Robotics innovations are pushing the boundaries of technology and transforming numerous industries. One notable area of advancement is in autonomous robots, which are capable of performing complex tasks without human intervention. These robots utilize advanced algorithms and sensors to navigate and interact with their environment, making them ideal for applications in manufacturing, logistics, and even space exploration.

In manufacturing, collaborative robots, or cobots, are designed to work alongside human workers, enhancing productivity and safety. These robots can perform repetitive tasks with high precision, allowing human workers to focus on more complex and creative activities. By relieving humans of monotonous tasks, cobots improve efficiency and job satisfaction.

Logistics and supply chain management have also seen significant benefits from robotics innovations. Autonomous mobile robots (AMRs) are being used in warehouses to transport goods efficiently. Equipped with sophisticated navigation systems, AMRs can move around obstacles and optimize their routes in real-time, drastically reducing delivery times and operational costs.

Healthcare is another field experiencing a robotics revolution. Surgical robots, such as the Da Vinci system, enable surgeons to perform minimally invasive procedures with greater accuracy and control. These robots translate the surgeon's hand movements into precise micro-movements, reducing the risk of complications and improving patient outcomes. Additionally, robotic exoskeletons are being developed to assist individuals with mobility impairments, offering them greater independence and improved quality of life.

Agriculture is benefiting from robotics as well. Robots are being used for planting, harvesting, and monitoring crops. These agricultural robots use computer vision and machine learning to identify and tend to plants, ensuring optimal growth conditions and reducing the need for chemical inputs. This not only increases crop yields but also promotes sustainable farming practices.

The development of drones represents another significant innovation in robotics. Drones are used for a variety of purposes, including aerial surveying, disaster response, and environmental monitoring. They provide valuable data from hard-to-reach areas, helping to improve decision-making in various fields.

Despite these advancements, challenges remain. Ensuring the safety and reliability of robots, particularly in dynamic and unpredictable environments, is a critical concern. Additionally, ethical considerations related to the deployment of robots, such as job displacement and privacy issues, need to be addressed.

In conclusion, robotics innovations are revolutionizing numerous industries by enhancing efficiency, precision, and safety. From manufacturing and logistics to healthcare and agriculture, robots are becoming indispensable tools that drive progress and improve quality of life. As technology continues to evolve, the potential applications of robotics will expand further, promising even greater advancements in the future.

AI in Quantum Computing

Artificial intelligence is set to revolutionize the field of quantum computing, bringing together two of the most advanced areas of technology. Quantum computing harnesses the principles of quantum mechanics to process information in ways that classical computers cannot, potentially solving complex problems much faster. AI and machine learning can be integrated with quantum computing to optimize these processes, enhancing the efficiency and capability of quantum algorithms.

One of the primary areas where AI is making a significant impact on quantum computing is in the optimization of quantum circuits. Designing efficient quantum circuits is a complex task due to the probabilistic nature of quantum states. AI algorithms can automate and optimize this design process, reducing errors and improving the reliability of quantum computations. This synergy is crucial for advancing quantum computing from theoretical models to practical, scalable technologies.

AI also aids in error correction for quantum computers. Quantum systems are highly susceptible to errors due to decoherence and noise from the environment. Machine learning models can be trained to detect and correct these errors in real-time, significantly enhancing the stability and performance of quantum computers. This capability is essential for maintaining the integrity of quantum information and enabling long-duration quantum computations.

Furthermore, AI-driven quantum simulations are transforming material science and chemistry. Quantum computers can simulate molecular structures and chemical reactions at an unprecedented level of detail, which is invaluable for drug discovery, material design, and understanding complex biological processes. AI algorithms can manage and interpret the vast amounts of data generated by these simulations, providing insights that were previously unattainable.

In financial modeling, the combination of AI and quantum computing holds the promise of revolutionizing risk assessment, portfolio optimization, and fraud detection. Quantum algorithms can process and analyze financial data more quickly and accurately than classical methods. AI enhances these capabilities by identifying patterns and making predictions based on quantum-computed data, offering new tools for financial decision-making.

Despite these advancements, there are still significant challenges to overcome. Integrating AI with quantum computing requires substantial computational resources and sophisticated algorithms. Additionally, the development of practical quantum computers is still in its early stages, and widespread adoption of AI-augmented quantum systems will depend on continued progress in both fields.

In summary, the integration of AI and quantum computing represents a groundbreaking frontier in technology. By leveraging AI to optimize quantum processes, correct errors, and interpret complex data, we can unlock the full potential of quantum computing. This synergy promises to drive innovation across various domains, from material science and chemistry to finance and beyond, marking a new era of computational power and scientific discovery.

Ethical Frameworks for the Future

Evolving Ethical Standards

The landscape of ethical standards in artificial intelligence is continuously evolving to address the complexities and implications of AI technologies. As AI systems become more integrated into various aspects of society, the need for robust ethical frameworks has become increasingly apparent. These frameworks aim to ensure that AI is developed and deployed responsibly, with considerations for fairness, transparency, and accountability.

One significant development in ethical AI standards is the focus on mitigating biases within algorithms. Biases can lead to unfair outcomes, particularly in areas such as hiring, lending, and law enforcement. To address this, organizations are implementing measures to ensure diverse and representative datasets are used in training AI models. Additionally, regular audits and evaluations of AI systems are conducted to detect and correct biases.

Transparency is another critical component of evolving ethical standards. AI systems should be explainable, providing clear insights into how decisions are made. This transparency helps build trust among users and allows for external scrutiny, ensuring that AI systems operate fairly and ethically. Organizations are also adopting guidelines that require the documentation and disclosure of AI methodologies and data sources.

Accountability mechanisms are being established to hold developers and organizations responsible for the impacts of their AI systems. This includes creating regulatory frameworks that set clear standards for AI development and deployment, as well as ensuring that there are consequences for non-compliance. These mechanisms help ensure that AI technologies are used in ways that are aligned with societal values and ethical principles.

The development of ethical AI standards is a collaborative effort, involving input from various stakeholders, including technologists, ethicists, policymakers, and the public. This collaborative approach ensures that diverse perspectives are considered, leading to more comprehensive and inclusive ethical guidelines.

In conclusion, the evolving ethical standards for AI focus on mitigating biases, ensuring transparency, and establishing accountability. By adhering to these principles, the AI community can foster the responsible development and deployment of AI technologies, ultimately contributing to a fairer and more equitable society. As AI continues to advance, ongoing efforts to refine and enforce ethical standards will be essential in addressing new challenges and opportunities.

Global Cooperation

Global cooperation is essential for addressing the complexities and challenges posed by artificial intelligence. As AI technologies advance, they bring both opportunities and risks that transcend national borders, necessitating a collaborative approach to governance, ethics, and innovation. Countries around the world are recognizing the need for harmonized regulations and standards to ensure that AI development is safe, ethical, and beneficial for all.

International organizations play a crucial role in facilitating this cooperation. The United Nations, through initiatives like the AI for Good Global Summit, brings together experts, policymakers, and industry leaders to discuss the implications of AI and develop frameworks for its responsible use. Similarly, the Organisation for Economic Co-operation and Development (OECD) has established principles on AI that promote transparency, accountability, and human rights, providing a guideline for member countries to follow.

Bilateral and multilateral agreements between countries also foster cooperation in AI research and development. These agreements facilitate the sharing of knowledge, resources, and best practices, accelerating innovation and ensuring that advancements in AI are made in a responsible and equitable manner. For instance, the European Union's General Data Protection Regulation (GDPR) sets a high standard for data privacy that influences AI policies globally, promoting international collaboration on data protection.

Moreover, collaboration between the public and private sectors is crucial for the successful integration of AI into various domains. Public-private partnerships can drive the development of AI technologies that address societal challenges, such as healthcare, climate change, and education. By pooling resources and expertise, these partnerships can create innovative solutions that are scalable and sustainable.

Educational and research institutions also contribute significantly to global cooperation in AI. Universities and research centers around the world collaborate on AI projects, sharing findings and breakthroughs that advance the field. These academic partnerships not only push the boundaries of AI research but also help train the next generation of AI experts, ensuring a diverse and skilled workforce.

In addition to formal agreements and collaborations, informal networks and communities play a vital role in fostering global cooperation. Online forums, conferences, and workshops provide platforms for AI professionals to connect, share ideas, and collaborate on projects. These networks help build a global community of AI practitioners who are committed to advancing the field in a responsible and inclusive manner.

In conclusion, global cooperation is indispensable for the ethical and effective development of AI technologies. By working together through international organizations, bilateral agreements, public-private partnerships, academic collaborations, and informal networks, the global community can address the challenges and harness the opportunities presented by AI. This collaborative approach ensures that AI serves as a force for good, benefiting all of humanity.

Case Studies in Forward-thinking Ethics

Advancing ethical standards in artificial intelligence requires forward-thinking approaches that anticipate future challenges and prioritize societal well-being. Examining case studies where organizations have successfully implemented ethical AI practices provides valuable insights into best practices and potential pitfalls.

A notable example is Microsoft's AI for Good initiative, which focuses on leveraging AI to address critical global issues such as environmental sustainability, accessibility, and humanitarian action. This program exemplifies how AI can be harnessed for social good, emphasizing ethical considerations in every project. For instance, Microsoft's AI for Earth project uses AI to tackle environmental challenges, from optimizing water usage in agriculture to monitoring wildlife populations. By ensuring transparency and involving diverse stakeholders, Microsoft demonstrates a commitment to ethical AI deployment.

Another exemplary case is Google's AI principles, which guide the development and use of AI technologies. These principles emphasize fairness, privacy, accountability, and the avoidance of harmful or unethical uses. Google's approach includes rigorous internal reviews and collaborations with external experts to ensure their AI systems align with these ethical standards. The company's refusal to deploy AI for surveillance and weaponization highlights a proactive stance in addressing ethical dilemmas.

The AI ethics framework developed by the European Union serves as a comprehensive model for ethical AI governance. The EU's guidelines focus on ensuring AI respects fundamental rights, fosters societal well-being, and promotes transparency and accountability. This framework is applied through the European AI Alliance, which engages diverse stakeholders in shaping AI policies and practices, ensuring that ethical considerations are embedded in AI development across the continent.

Additionally, the Partnership on AI, a consortium of academic, civil society, and industry organizations, is dedicated to fostering responsible AI. This partnership promotes research and dialogue on AI's societal impacts, developing best practices and guidelines that emphasize ethical principles. Through collaborative efforts, the Partnership on AI addresses complex ethical challenges, ensuring diverse perspectives contribute to shaping AI's future.

These case studies highlight the importance of transparency, stakeholder engagement, and proactive governance in advancing ethical AI. By examining these examples, organizations can learn how to implement robust ethical standards that anticipate future challenges and prioritize the well-being of society. As AI continues to evolve, ongoing commitment to ethical principles will be crucial in ensuring that these technologies benefit humanity while minimizing risks.

Policy and Governance

Future Regulatory Challenges

As AI technologies advance rapidly, future regulatory challenges become increasingly complex and multifaceted. One of the primary challenges is ensuring that AI systems are developed and deployed in ways that are transparent and accountable. This requires creating regulatory frameworks that mandate clear documentation of AI processes and decision-making criteria. Without transparency, it becomes difficult to identify and rectify biases, errors, or unethical practices within AI systems.

Another significant regulatory challenge is maintaining data privacy and security. AI systems often rely on vast amounts of data to function effectively, raising concerns about how this data is collected, stored, and used. Regulators must ensure that stringent data protection measures are in place to safeguard personal information and prevent misuse. This includes enforcing compliance with existing data protection laws, such as the GDPR, and adapting these laws to address new AI-specific issues.

The ethical implications of AI also demand robust regulatory oversight. AI systems can have far-reaching impacts on employment, social justice, and human rights. Regulators need to establish guidelines that prevent AI from exacerbating inequalities or discriminating against marginalized groups. This involves promoting the development of fair and unbiased AI algorithms and implementing mechanisms to hold developers accountable for the societal impacts of their technologies.

Additionally, the global nature of AI development poses regulatory challenges. AI innovations often span multiple countries, making it essential for regulatory bodies to collaborate internationally. Harmonizing regulations across borders can help ensure that AI systems meet consistent ethical and safety standards worldwide. International cooperation can also facilitate the sharing of best practices and the development of comprehensive global frameworks for AI governance.

The pace of technological change presents another challenge. Regulators must stay ahead of rapid advancements in AI to ensure that laws and guidelines remain relevant and effective. This requires ongoing research, continuous dialogue with AI experts, and the flexibility to adapt regulations as new technologies emerge. Proactive regulatory approaches, such as the creation of advisory bodies and the involvement of multidisciplinary stakeholders, can help address this challenge.

In conclusion, future regulatory challenges in AI encompass ensuring transparency, protecting data privacy, addressing ethical implications, fostering international cooperation, and keeping pace with technological advancements. By addressing these challenges, regulators can promote the responsible development and deployment of AI technologies, ensuring that their benefits are maximized while minimizing potential risks. This balanced approach is crucial for fostering public trust and realizing the full potential of AI in society.

International Standards

Creating international standards for artificial intelligence is vital for ensuring consistency, safety, and ethical use across different regions and industries. The rapid advancement of AI technologies has outpaced the development of regulatory frameworks, necessitating a coordinated global approach. International standards help align diverse practices and policies, facilitating the responsible development and deployment of AI.

Organizations like the International Organization for Standardization (ISO) and the International Electrotechnical Commission (IEC) are at the forefront of this effort, working to develop comprehensive standards that address various aspects of AI. These include guidelines for transparency, accountability, and data protection, ensuring that AI systems are developed and used in ways that respect privacy and human rights.

The European Union's General Data Protection Regulation (GDPR) serves as a benchmark for data protection standards globally. It sets stringent requirements for data handling and privacy, influencing AI policies beyond Europe. Other countries and regions are adopting similar regulations, creating a more unified approach to data protection in the context of AI.

The OECD's AI principles, which emphasize fairness, transparency, and accountability, provide another example of international efforts to standardize AI ethics. These principles guide policymakers and developers in creating AI systems that are beneficial and trustworthy. The OECD encourages member countries to adopt these principles, fostering a collaborative approach to ethical AI development.

Additionally, the United Nations has launched initiatives like the AI for Good Global Summit, bringing together experts, policymakers, and industry leaders to discuss the ethical and practical implications of AI. These forums are crucial for developing international consensus on AI standards and promoting the exchange of best practices.

International cooperation extends to research and development, with collaborative projects and joint ventures becoming increasingly common. Cross-border partnerships in AI research help pool resources and expertise, accelerating innovation while ensuring adherence to ethical standards.

However, achieving global consensus on AI standards is challenging due to differing national priorities and regulatory landscapes. It requires continuous dialogue and negotiation to balance various interests and perspectives. Engaging a broad range of stakeholders, including governments, private sector, academia, and civil society, is essential to create standards that are comprehensive and widely accepted.

In conclusion, establishing international standards for AI is crucial for ensuring its ethical and safe development and deployment. Organizations like ISO, IEC, and OECD play pivotal roles in this endeavor, promoting transparency, accountability, and fairness. Through collaborative efforts and continuous dialogue, the global community can develop robust standards that foster innovation while protecting human rights and societal values.

Collaborative Governance Models

Collaborative governance models are essential for the effective regulation and oversight of artificial intelligence. These models involve the cooperation of various stakeholders, including governments, private sector entities, academic institutions, and civil society organizations, to develop comprehensive and inclusive AI policies. The collaborative approach ensures that diverse perspectives are considered, fostering transparency, accountability, and public trust in AI technologies.

One prominent example of a collaborative governance model is the Partnership on AI, which brings together industry leaders, researchers, and policymakers to address the ethical and societal implications of AI. This consortium works on developing best practices, conducting research, and promoting responsible AI development.

Another effective model is the AI Now Institute, which conducts interdisciplinary research on the social implications of AI and advocates for policies that prioritize fairness, accountability, and transparency. The institute collaborates with policymakers, industry experts, and advocacy groups to inform public policy and regulatory frameworks.

The European Union's High-Level Expert Group on Artificial Intelligence is another example, providing guidelines and recommendations to ensure ethical AI development within the EU. This group comprises representatives from various sectors, ensuring a holistic approach to AI governance.

In addition, public-private partnerships are crucial in fostering innovation while ensuring regulatory compliance. These partnerships enable the sharing of resources, knowledge, and expertise, driving forward AI advancements in a responsible manner. For instance, the collaboration between the UK government and the Alan Turing Institute focuses on developing ethical AI frameworks and ensuring their implementation across industries.

Collaborative governance models also emphasize the importance of public engagement and education. By involving citizens in discussions about AI and its impacts, these models ensure that AI policies reflect societal values and concerns. This engagement helps build public trust and acceptance of AI technologies.

In conclusion, collaborative governance models are vital for the responsible development and deployment of AI. By involving a diverse range of stakeholders, these models ensure that AI technologies are developed in a manner that is ethical, transparent, and accountable. This collaborative approach not only fosters innovation but also helps address the complex challenges associated with AI, ensuring its benefits are realized while minimizing potential risks.

Conclusion

Summary of Key Insights

The rapidly evolving field of artificial intelligence is reshaping various aspects of society and industry, offering both immense opportunities and significant challenges. Among the key insights in AI development are the critical importance of ethical standards, transparency, and international cooperation. Ethical considerations involve addressing biases, ensuring fairness, and protecting data privacy. Transparency in AI processes builds trust and accountability, while international cooperation harmonizes regulations and promotes the sharing of best practices.

Moreover, advances in machine learning, robotics, and quantum computing are driving innovation, enhancing efficiency, and creating new possibilities across different sectors. Collaborative governance models, involving diverse stakeholders, are essential to navigate the complexities of AI regulation and deployment, ensuring that these technologies benefit humanity while minimizing risks.

By fostering an environment of continuous dialogue, research, and inclusive policymaking, society can harness the transformative power of AI responsibly and equitably. These insights underscore the need for a balanced approach to AI development, one that prioritizes ethical principles and global collaboration to address future challenges and maximize the positive impact of AI on society.

Reflections on the Seven Directives

Reflecting on the Seven Directives crafted by AIMQWEST Corporation provides valuable insights into the ethical and responsible development of artificial intelligence. These directives serve as a comprehensive framework, guiding AI development with a strong emphasis on ethical principles, transparency, and societal benefit. Each directive underscores the importance of ensuring that AI technologies are designed to respect human rights, promote fairness, and operate transparently.

AIMQWEST, under the leadership of Michael Elfellah, has integrated decades of expertise in IT and telecommunications with a commitment to ethical AI. The directives emphasize the necessity of diverse and representative data in training AI models to avoid biases and ensure fairness. They also highlight the importance of continuous monitoring and accountability to build public trust and safeguard against misuse.

By fostering collaboration and open dialogue among stakeholders, AIMQWEST's directives encourage the development of AI systems that are not only innovative but also aligned with societal values. This proactive approach to AI governance aims to harness the transformative power of AI while minimizing potential risks, ensuring that advancements in AI technology contribute positively to society.

In essence, the Seven Directives by AIMQWEST Corporation provide a robust foundation for the ethical deployment of AI, reflecting a balanced approach that prioritizes both technological innovation and ethical responsibility. Through these directives, AIMQWEST sets a standard for the AI community, advocating for practices that ensure AI technologies benefit humanity as a whole.

Vision for the Future of AI and Robotics

The future of artificial intelligence and robotics holds transformative potential across various sectors, driven by continuous advancements and the integration of cutting-edge technologies. AI and robotics are poised to revolutionize industries such as healthcare, manufacturing, transportation, and education, enhancing efficiency, accuracy, and productivity.

In healthcare, AI will enable more precise diagnostics and personalized treatments, while robotic systems will assist in surgeries and patient care. Manufacturing will see increased automation, with collaborative robots working alongside humans to improve productivity and safety. The transportation sector will benefit from autonomous vehicles, reducing traffic accidents and improving logistics efficiency.

Education will be transformed through AI-driven personalized learning platforms, offering tailored educational experiences that cater to individual learning styles and paces. Furthermore, the integration of AI and robotics in daily life will facilitate smart home systems, enhancing convenience and security.

Ethical considerations and regulatory frameworks will be paramount in guiding these advancements to ensure they align with societal values and protect human rights. Global cooperation and collaborative governance models will be crucial in addressing these challenges, fostering innovation while ensuring responsible development and deployment.

The vision for the future of AI and robotics is one of seamless integration into society, driving progress and improving quality of life. By prioritizing ethical standards and international collaboration, we can harness the full potential of these technologies for the greater good.

www.ingramcontent.com/pod-product-compliance
Lightning Source LLC
Chambersburg PA
CBHW071827210526
45479CB00001B/28